W E Fitzgibbon (III) & H F Walker (Editors)

University of Houston

Nonlinear diffusion

Pitman

LONDON · SAN FRANCISCO · MELBOURNE

PITMAN PUBLISHING LIMITED
39 Parker Street, London WC2B 5PB

FEARON-PITMAN INC.
6 Davis Drive, Belmont, California 94002, USA

Associated Companies
Copp Clark Ltd, Toronto
Pitman Publishing Co. SA (Pty) Ltd, Johannesburg
Pitman Publishing New Zealand Ltd, Wellington
Pitman Publishing Pty Ltd, Melbourne
Sir Isaac Pitman Ltd, Nairobi

First published 1977

AMS Subject Classifications: (main) 35-02, 35K55
(subsidiary) 35G20, 35B99, 92A15, 76R99

Library of Congress Cataloging in Publication Data

Main entry under title:
Nonlinear diffusion.

(Research notes in mathematics; 14)
1. Diffusion processes. 2. Differential equations, Parabolic. I. Walker, Homer Franklin, 1943- II. Fitzgibbon, William Edward, 1945- III. Series.
QA274.75.N66 519.2'33 77-8501
ISBN 0-273-01066-2

Reproduced and printed by photolithography
in Great Britain by Biddles of Guildford

000315391

Nonlinear diffusion

[redacted]

Preface

In recent years, nonlinear diffusion equations have attracted increasing interest from the mathematical community, a fact which reflects growth in the range and importance of their applications as well as recent contributions to the understanding of the mathematical behavior of their solutions. Such equations arise in the mathematical modeling of a number of social, biological and physical phenomena occurring in such diverse areas as population dynamics, population genetics, epidemiology, nerve impulse studies, electrical transmission line analysis, chemical reactor theory, neutron transport theory and the study of gas flow in porous media. The study of the solutions of such equations has proved to be a rich and rewarding area of mathematical endeavor, yielding not only valuable insights into the phenomena being modeled but also mathematical results which often depart surprisingly from the results of the linear theory. Directions taken by this study include both those unique to this area, such as threshold analysis, and those more often associated with other types of equations, such as inquiries concerning the existence of traveling waves and finite speeds of signal propagation.

With the sponsorship and support of the N.S.F.-C.B.M.S. regional conference program, a five-day regional conference on the subject of nonlinear diffusion equations was held at the University of Houston in June, 1976. To our knowledge, this conference was the first conference exclusively devoted to nonlinear diffusion to be held in the United States. Participants were drawn from across the United States with the majority coming from the Southern and Southwestern geographical regions. The principal speaker for the conference was Donald G. Aronson of the University of Minnesota. The ten one-hour lectures delivered by Professor Aronson are to be published by S.I.A.M. in the C.B.M.S. Regional Conference Series in Applied Mathematics.

Professor Aronson's lectures were complemented by invited and contributed talks from the conference participants. This collection contains most of these talks as well as several additional contributed papers. Because of the timeliness of the subject and the lack of comprehensive and current treatments of it, as well as the expertise and range of interests of the authors, we feel that this collection should be of considerable value to the mathematical community.

The editors would like to take the opportunity to thank the conference participants for making it a pleasant and mathematically exciting event. Also, the conference could not have been successful without the enthusiastic support of the staffs of the Mathematics Department, the College of Natural Sciences and Mathematics, and the Cullen College of Engineering of the University of Houston. Particular credit is due D. Brown, R. Byrd, G. Graham, C. Horgan, and L. Wheeler for assisting in the organization and management of the conference. Finally, special thanks must be extended to Rachael Kidd for her diligent work in the preparation of this typescript.

Houston, Texas
March, 1977

W. E. Fitzgibbon
H. F. Walker

Contents

D. G. Aronson:

The Asymptotic Speed of Propagation of a Simple Epidemic 1

J. R. Cannon and R. E. Ewing:

Galerkin Procedures for Systems of Parabolic Partial Differential Equations Related to the Transmission of Nerve Impulses 24

E. D. Conway and J. A. Smoller:

Diffusion and the Classical Ecological Interactions: Asymptotics 53

J. W. Evans:

Transition Behavior at the Slow and Fast Impulses 70

P. C. Fife:

Stationary Patterns for Reaction-Diffusion Equations 81

D. Henry:

Gradient Flows Defined by Parabolic Equations 122

N. J. Kopell:

Waves, Shocks and Target Patterns in an Oscillating Chemical Reagent 129

R. M. Miura:

A Nonlinear WKB Method and Slowly-Modulated Oscillations in Nonlinear Diffusion Processes 155

P. Nelson:

Subcriticality for Submultiplying Steady-State Neutron Diffusion 171

J. Rinzel:

Repetitive Nerve Impulse Propagation: Numerical Results and Methods 186

M. E. Schonbek:

Some Results on the FitzHugh-Nagumo Equations 213

A. D. Snider and D. L. Akins:

Calculation of Transients for Some Nonlinear Diffusion Phenomena 218

D G ARONSON

The asymptotic speed of propagation of a simple epidemic

ABSTRACT

D. G. Kendall proposed a model for a simple epidemic with removal but without recovery in a spatially stratified population. Kendall analyzed an approximate version of his model in which the dispersion operators are replaced by standard diffusion operators. He proved the existence of traveling waves with all velocities greater than or equal to a certain minimum. In subsequent work, I proved that Kendall's minimal velocity is the asymptotic speed of propagation of disturbances from the rest state. In this paper I extend this result to Kendall's original system of equations.

* * * *

In 1965 D. G. Kendall [7] initiated the mathematical study of the spread of a disease in a spatially inhomogeneous population.* The model he considered is an adaptation of a model proposed in 1927 by Kermack and McKendrick [8] for epidemics in spatially homogeneous populations. The basic model describes an epidemic in which there is removal but no recovery. The mechanism for the spread of an infection is of the mass action type, that is, the rate of infection of susceptible individuals is proportional to the product of the density of susceptibles and the density of the infectious individuals.

---

* Reading Kendall's paper is a rewarding experience. It contains an elegant presentation of an interesting piece of mathematics and a very sensible view of the role of mathematical modeling in the biological sciences.

Kendall retained the form of the Kermack-McKendrick equations, but introduced spatial dependence by replacing the density of infectious individuals in the mass action terms by a spatially averaged density. In mathematical terms, the Kermack-McKendrick model involves a system of ordinary differential equations which is replaced by a system of integro-differential equations in Kendall's model.

Given a spatially inhomogeneous epidemic model it is quite natural to look for traveling wave solutions. Kendall did this for an approximate version of his model in which the nonlocal operation of averaging the density of infectious individuals is replaced by a local differential operator. In this diffusion approximation the system of integro-differential equations is replaced by a system of partial differential equations involving a nonlinear diffusion. Roughly speaking, Kendall proved the existence of a positive number c^* such that his model admits traveling wave solutions of all speeds $c \geq c^*$ and no traveling wave solutions with speeds less than c^*. Mollison studied Kendall's original model in the special case in which there are no removals. With this assumption the system of integro-differential equations reduces to a single equation. For a particular choice of the averaging kernel Mollison [11] proves the analog of Kendall's result. For general averaging kernels [10] he obtains a criterion for the existence of wave speeds and wave form bounds for propagating solutions. Quite recently, Atkinson and Reuter [4] analyzed the full Kendall model for a general class of averaging kernels. They obtain a criterion for the existence of a critical speed $c^* > 0$ and the existence of traveling waves of all speeds $c > c^*$. Daniels [5] discusses techniques for approximating traveling wave solutions.

Minimal wave speeds analogous to those found by Kendall, Mollison, and Atkinson and Reuter also occur in the classical work of Fisher [6] and

Kolmogoroff, Petrovsky and Piscounoff [9] concerning the advance of advantageous genes. Weinberger and I have shown [2], [3] that the minimal wave speed is the asymptotic speed of propagation of disturbances from the rest state for Fisher's equation and it is natural to conjecture that an analogous result holds for Kendall's epidemic model. In my lecture notes [1] I have verified this conjecture for the diffusion approximation to Kendall's model. In this paper I will prove the corresponding result for the original model.

Consider a population distributed on an infinite one dimensional habitat and suppose the population is divided into three classes:

(i) susceptible individuals with local density u

(ii) infected individuals with local density v

(iii) removed individuals with local density w.

Here u, v, and w are functions of the location x in the habitat and the time t, with $u + v + w$ independent of t. All infected individuals are assumed to be infectious and the rate of infection is given by

$$\gamma u(k*v) ,$$

where $\gamma > 0$ is a constant and

$$(k*v)(x,t) = \int_R v(x - y,t)k(y)dy .$$

Here k is a nonnegative kernel with

$$\int_R k(y)dy = 1 .$$

Removed individuals can be regarded as being either immune or dead and the rate of removal is assumed to be ρv, where $\rho > 0$ is constant. With this notation, Kendall's equations are

$$
\left.\begin{aligned}
u_t &= -\gamma u(k*v) \\
v_t &= \gamma u(k*v) - \rho v \\
w_t &= \rho v
\end{aligned}\right\} \quad (1)
$$

where the subscript denotes partial differentiation. The equation for w is, of course, superfluous since w can be computed once v is known.

A rest state for the system (1) is given by $u \equiv \sigma$, $v \equiv w \equiv 0$ where $\sigma > 0$ is constant. I will be interested in the asymptotic behavior of solutions of (1) which correspond to perturbations from this rest state. Specifically, I will consider the initial value problem for the system (1) in $R \times R^+$ with initial data

$$
u(x,0) \equiv \sigma \ , \quad v(x,0) = v_0(x) \ , \quad w(x,0) \equiv 0 \quad \text{in} \quad R
$$

where

$$
\left.\begin{aligned}
&v_0 \in C(R) \ \text{ is such that } \ v_0 \geq 0 \ , \ v_0 \not\equiv 0 \ , \ \text{and} \\
&v_0(x) \equiv 0 \ \text{ in } \ [x_0, +\infty) \ \text{ for some } \ x_0 \in R.
\end{aligned}\right\} \quad (2)
$$

By rescaling u, v, w, and t the problem I will consider can be written in the form

$$
\left.\begin{aligned}
&\left.\begin{aligned}
u_t &= -u(k*v) \\
v_t &= u(k*v) - \lambda v \\
w_t &= \lambda v
\end{aligned}\right\} \quad \text{in } R \times R^+ \\
&u(x,0) = 1 \ , \quad v(x,0) = v_0(x) \ , \quad w(x,0) = 0 \quad \text{in } R
\end{aligned}\right\} \quad (3)
$$

where $\lambda = \rho/\gamma\sigma$ and v_0 satisfies (2) .

Suppose that u, v, w is a solution of problem (3). Then since

$$
(u + v + w)_t = 0
$$

it follows that

$$u + \frac{1}{\lambda} w_t + w = 1 + v_0(x) .$$

From the first and third equations in (3), u is given by

$$u(x,t) = \exp \{-\int_0^t (k*v)(x,\tau)d\tau\}$$

and w is given by

$$w(x,t) = \lambda \int_0^t v(x,\tau)d\tau .$$

Therefore

$$u(x,t) = \exp\{-\int_R k(x - y)[\int_0^t v(y,\tau)d\tau]dy\} = \exp\{-\frac{1}{\lambda}(k*w)(x,t)\}$$

and

$$e^{-\frac{1}{\lambda} k*w} + \frac{1}{\lambda} w_t + w = 1 + v_0(x) .$$

Thus if u, v, w is a solution of (3), then w satisfies

$$\left.\begin{aligned} &w_t = -\lambda w + \lambda\{1 - e^{-\frac{1}{\lambda} k*w}\} + \lambda v_0(x) \quad \text{in} \quad R \times R^+ \\ &w(x,0) = 0 \quad \text{in} \quad R. \end{aligned}\right\} \qquad (4)$$

Conversely, if w is a solution of problem (4) then u, v, w is a solution of problem (3) where

$$u = e^{-\frac{1}{\lambda}(k*w)}$$

and

$$v = -w + \{1 - e^{-\frac{1}{\lambda} k*w}\} + v_0 .$$

I will restrict my attention to problem (4).

In order to get any results it is necessary to make some assumptions about the averaging kernel k. Specifically, I assume:

(K1) k is a nonnegative even function defined in R with

$$\int_R k(y)dy = 1 .$$

(K2) There exists a $\nu \in (0, +\infty]$ such that

$$\int_R e^{\mu y} k(y)dy < +\infty \quad \text{for all} \quad \mu \in [0,\nu) .$$

(K3) Define

$$A_\lambda(\mu) \equiv \frac{1}{\mu}\{ \int_R e^{\mu y}k(y)dy - \lambda\} .$$

For each $\lambda < 1$ there exists a $\mu^* = \mu^*(\lambda) \in (0,\nu)$ such that

$$0 < c^* \equiv A_\lambda(\mu^*) = \inf\{A_\lambda(\mu) : 0 < \mu < \nu\},$$

$$A_\lambda'(\mu) < 0 \quad \text{in} \quad (0,\mu^*) \quad ,$$

and

$$A_\lambda'(\mu) > 0 \quad \text{in} \quad (\mu^*,\nu) .$$

(K4) For each $\bar{\mu} \in (0,\nu)$ there exists an $r = r(\bar{\mu}) \in R^+$ such that

$$e^{\mu x}k(x) = \min\{e^{\mu y}k(y) : y \in [0,x]\}$$

for all $\mu \in [0,\bar{\mu}]$ and $x \geq r(\bar{\mu})$.

I will refer to these assumptions collectively as (K) . The assumptions (K1) and (K4) are somewhat stronger than necessary. For example, one can drop the evenness of k. On the other hand, Mollison [10] obtains (K2) as a necessary condition for a finite speed of propagation in his work. Moreover, $c^* = A_\lambda(\mu^*)$ agrees with the minimal wave speed obtained by Atkinson and Reuter [4].

The first result corresponds to the Kermack-McKendrick-Kendall Threshold Theorem (cf. [7]). Roughly speaking, it says that an initial infection (described by v_0) does not propagate if $\lambda \geq 1$, that is, if the initial density of susceptibles is too low or if the removal rate is too high.

Theorem 1. Let $w(x,t)$ denote the solution of problem (4), where v_0 satisfies (2) and the kernel k satisfies (K). If $\lambda \geq 1$ then for every $x \in R$ and $c \in R^+$

$$\lim_{t\uparrow\infty} w(x + ct,t) = 0 .$$

The next result shows that the situation is quite different for $\lambda \in (0,1)$.

Theorem 2. Let $w(x,t)$ denote the solution of problem (4), where v_0 satisfies (2), k satisfies (K), and $\lambda \in (0,1)$.

(a) If $c > c^*$ then for every $x \in R$

$$\lim_{t\uparrow\infty} w(x + ct,t) = 0.$$

(b) If $0 < c < c^*$ then for every $x \in R$

$$\lim_{t\uparrow\infty} w(x + ct,t) = \alpha(\lambda) ,$$

where $\alpha = \alpha(\lambda)$ is the unique solution in $(0,1)$ of

$$1 - \alpha = e^{-\alpha/\lambda} .$$

Roughly speaking, Theorem 2 says that if you travel toward $+\infty$ from any point in R then you will outrun the infection if your speed exceeds the minimal speed c^*, but the infection will overtake you if your speed is less than c^*. Note that $w \equiv \alpha(\lambda)$ is an equilibrium solution of the homogeneous equation

$$w_t = -\lambda w + \lambda\{1 - e^{-\frac{1}{\lambda} k*w}\} .$$

Moreover, although I will not prove it here, the conclusion of Theorem 2(b) also holds for $c = 0$.

The proofs of Theorem 1 and 2(a) as well as the estimate for the limit superior in Theorem 2(b) are based on certain a priori estimates which are given below in Propositions 1 and 2. The estimate for the limit inferior in Theorem 2(b) depends on the construction of a subsolution for problem (4). The ideas involved in this construction evolved in a series of discussions with H. F. Weinberger on nonlocal discrete time models in population genetics (cf. [12]) and I gratefully acknowledge my debt to him.

To prove Theorems 1 and 2 it is most convenient to work in a coordinate frame which moves with speed c. Let $w(x,t)$ be the solution of the initial value problem (4). For any $c \geq 0$ set $x = \xi + ct$ and define

$$W(\xi,t) \equiv w(\xi + ct,t) .$$

Then W satisfies

$$\left.\begin{array}{l} W_t = Q_c[W] + \lambda v_0(\xi + ct) \quad \text{in} \quad R \times R^+ \\ W(\xi,0) = 0 \quad \text{in} \quad R , \end{array}\right\} \qquad (5)$$

where

$$Q_c[W] \equiv cW_\xi - \lambda W + E[W]$$

and

$$E[W] = \lambda\{1 - e^{-\frac{1}{\lambda} k*W}\} .$$

It is not difficult to show that for any bounded continuous functions F and G the initial value problem

$$W_t = Q_c[W] + F(\xi,t) \quad \text{in} \quad R \times R^+$$

$$W(\xi,0) = G(\xi) \quad \text{in} \quad R$$

has a unique solution which is bounded in $R \times [0,T]$ for any $T \in R^+$. Moreover, the solution depends continuously on its initial values in $R \times [0,T]$ for any $T \in R^+$. To prove this one writes the initial value problem as an integral equation which can then be solved by the usual method of successive approximations.

The main tool in the proofs of Theorems 1 and 2 is the following comparison result. Since I will have to apply it in several different situations, I will state it in a rather general form. Let E denote a mapping of the continuous functions on $R \times [0, +\infty)$ into themselves such that, for each fixed t, $0 \le U'(\xi,t) \le U(\xi,t)$ in R implies

$$0 \le E[U'](\xi,t) \le E[U](\xi,t) \quad \text{in} \quad R.$$

Moreover, let $C^1(R \times R^+)$ denote the class of functions U which are bounded and continuous in $R \times [0, +\infty)$, and which are such that U_ξ and U_t are continuous on a finite number of strips $(-\infty, \xi_0] \times R^+$, $[\xi_0, \xi_1] \times R^+, \ldots,$ $[\xi_N, +\infty) \times R^+$.

<u>Comparison Lemma</u>. Let U and V be $C^1(R \times R^+)$ functions such that

$$U_t - cU_\xi + \lambda U - E[U] \ge V_t - cV_\xi + \lambda V - E[V]$$

everywhere in $R \times R^+$ where the derivatives are continuous, and such that

$$U(\xi,0) > V(\xi,0) \quad \text{in} \quad R.$$

Then $U > V$ everywhere in $R \times [0, +\infty)$.

<u>Proof</u>. Suppose there exists a point $(\xi_0,t_0) \in R \times R^+$ such that $U > V$ in $R \times [0,t_0)$ and $U(\xi_0,t_0) = V(\xi_0,t_0)$. Set $Z \equiv U - V$. Then $Z(\xi,0) > 0$ in R and

$$Z_t - cZ \ge -\lambda Z + E[U] - E[V]$$

in $R \times R^+$ wherever the derivatives are defined. Integrate this differential inequality to obtain

$$Z(\xi,t) \geq Z(\xi + ct,0)e^{-\lambda t} + \int_0^t e^{-\lambda(t-\tau)}\{E[U] - E[V]\}(\xi + c(t - \tau),\tau)d\tau$$

in $R \times R^+$. Since $U > V$ in $R \times [0,t_0)$ it follows that $Z(\xi_0,t_0) > 0$ which contradicts $U(\xi_0,t_0) = V(\xi_0,t_0)$. Therefore $U > V$ in $R \times R^+$.

The condition $U(\xi,0) > V(\xi,0)$ in the Comparison Lemma is awkward and can be weakened whenever U or V is actually the solution of a well-posed initial value problem. For example, suppose that

$$U_t - cU_\xi + \lambda U - E[U] = F(\xi,t) \geq V_t - cV_\xi + \lambda V - E[V]$$

and

$$U(\xi,0) = G(\xi) \geq V(\xi,0) .$$

If U^ε denotes the solution of the initial value problem

$$U_t - cU_\xi + \lambda U - E[U] = F(\xi,t) \quad \text{in} \quad R \times R^+$$

$$U(\xi,0) = G(\xi) + \varepsilon \quad \text{in} \quad R$$

for arbitrary $\varepsilon > 0$, then, by the Comparison Lemma, $U^\varepsilon > V$ in $R \times R^+$. By hypothesis the initial value problem for U is well-posed so that $U^\varepsilon \to U \geq V$ as $\varepsilon \downarrow 0$. I will use this remark freely in the sequel without distinguishing it from the Comparison Lemma.

Proposition 1. Let F and G be bounded smooth functions and let $U(\xi,t)$ denote the solution of the initial value problem

$$U_t = Q_c[U] + F(\xi,t) \quad \text{in} \quad R \times R^+$$

$$U(\xi,0) = G(\xi) \quad \text{in} \quad R .$$

Then $U(\xi,t) > 0$ in $R \times R^+$ provided that $F \not\equiv 0$ or $G \not\equiv 0$.

Proof. Set $U^0 \equiv G$. For each integer $n \geq 1$ let U^n be defined by

$$U_t^n = cU_\xi^n - \lambda U^n + E[U^{n-1}] + F \quad \text{in} \quad R \times R^+$$

$$U^n(\xi,0) = G(\xi) \quad \text{in} \quad R.$$

By direct computation $U^n \geq U^{n-1}$ in $R \times R^+$ for all $n \geq 1$ and, by the Comparison Lemma, $U \geq U^n$ in $R \times R^+$ for all $n \geq 1$. If the support of either k or G or F (as a function of ξ for fixed t) is R then $U^1 > 0$ in $R \times R^+$ and the assertion follows. Suppose that $F \equiv 0$, $\operatorname{supp} G = [\xi_0, \xi_1]$ and $\operatorname{supp} k = [-\ell,\ell]$. Then for each fixed $t \in R^+$ the support of U^n as a function of ξ is $[\xi_0 - n(\ell + ct), \xi_1 + n\ell]$. Thus for each $(\xi,t) \in R \times R^+$, $U^n(\xi,t) > 0$ for all sufficiently large n. A similar argument gives the same conclusion in case $G \equiv 0$ and F has compact support as a function of ξ.

The proofs of Theorems 1 and 2(a) are based on the following estimate which is a consequence of the Comparison Lemma and Duhamel's Principle.

Proposition 2. Let W denote the solution of the initial value problem (5), where v_0 satisfies (2) and k satisfies (K). Then

$$0 \leq W(\xi,t) \leq \frac{m(\mu)e^{-\mu\xi}}{|\mu A_\lambda(\mu)|} \left| \exp[\mu\{A_\lambda(\mu) - c\}t] - \exp[-c\mu t]\right| \tag{6}$$

in $R \times R^+$ for every $\mu \in (0,\nu)$ such that $A_\lambda(\mu) \neq 0$, where

$$m(\mu) = \lambda \sup_R e^{\mu\xi} v_0(\xi)$$

and

$$A_\lambda(\mu) = \frac{1}{\mu}\{ \int_R e^{\mu\xi} k(\xi)d\xi - \lambda\} .$$

Proof. Let Z denote the solution of linear initial value problem

$$Z_t = L_c[Z] + \lambda v_0(\xi + ct) \quad \text{in} \quad \mathcal{R} \times \mathcal{R}^+$$

$$Z(\xi,0) = 0 \quad \text{in} \quad \mathcal{R},$$

where

$$L_c[Z] = cZ_\xi - \lambda Z + k*Z \; .$$

Then $Z(\xi,0) = W(\xi,0)$ in $\mathcal{R}$ and, since $\lambda(1 - e^{-a/\lambda}) \le a$,

$$Z_t - Q_c[Z] \ge Z_t - L_c[Z] = \lambda v_0 = W_t - Q_c[W]$$

in $\mathcal{R} \times \mathcal{R}^+$. Thus, by the Comparison Lemma, $Z \ge W$ in $\mathcal{R} \times \mathcal{R}^+$.

It is easily verified that

$$Z(\xi,t) = \int_0^t Z^*(\xi,t;\eta)d\eta$$

where $Z^*(\xi,t;\eta)$ satisfies

$$Z_t^* = L_c[Z^*] \quad \text{in} \quad \mathcal{R} \times (\eta, +\infty)$$

$$Z^*(\xi,\eta;\eta) = \lambda v_0(\xi + c\eta) \quad \text{in} \quad \mathcal{R}.$$

This representation for Z is usually called Duhamel's Principle.

For arbitrary $\mu \in (0,\nu)$ set

$$U(\xi,t;\eta) \equiv e^{\mu\xi}Z^*(\xi,t;\eta) \; .$$

Then

$$U_t = cU_\xi - (\lambda + c\mu)U - \ell*U \quad \text{in} \quad \mathcal{R} \times (\eta, +\infty) \tag{7}$$

$$U(\xi,\eta;\eta) = e^{-\mu c\eta}y_0(\xi + c\eta) \; ,$$

where $\ell(\xi) \equiv e^{\mu\xi}k(\xi)$ and $y_0(\xi) \equiv \lambda e^{\mu\xi}v_0(\xi)$. The function

$$V(t;\eta) \equiv m(\mu)\exp\{\mu A(\mu)(t-\eta) - c\mu t\}$$

is a solution of the equation (7) in $R \times (\eta, +\infty)$ with

$$V(\eta;\eta) = m(\mu)e^{-c\mu\eta} \geq U(\xi,\eta;\eta) \quad \text{in} \quad R.$$

Therefore, by the Comparison Lemma, $V(t;\eta) \geq U(\xi,t;\eta)$ in $R \times [\eta, +\infty)$. The estimate (6) follows from

$$0 \leq W(\xi,t) \leq e^{-\mu\xi}\int_0^t U(\xi,t;\eta)d\eta \leq e^{-\mu\xi}\int_0^t V(t;\eta)d\eta$$

by integration.

Proof of Theorem 1. It suffices to show that the right hand side of (6) tends to zero as $t \uparrow \infty$. Suppose first that $\lambda > 1$. Then since

$$\int_R e^{\mu\xi}k(\xi)d\xi \downarrow 1 \quad \text{as} \quad \mu \downarrow 0$$

it follows that $A_\lambda(\mu) \to -\infty$ as $\mu \downarrow 0$. If $\lambda = 1$ then by L'Hospital's rule and the evenness of k

$$\lim_{\mu\downarrow 0} A_1(\mu) = \lim_{\mu\downarrow 0}\int_R \xi e^{\mu\xi}k(\xi)d\xi = 0 .$$

Thus $A_\lambda(\mu) - c < 0$ for all sufficiently small $\mu > 0$ provided that $c > 0$ and $\lambda \geq 1$. Hence for suitably chosen μ the right hand side of (6) tends to zero as $t \uparrow \infty$.

Proof of Theorem 2(a). In view of (K3), for each $c > c^*$ there exists a value of $\mu \in (0,\mu^*)$ such that $c > A(\mu) > 0$. Thus the assertion follows immediately from Proposition 2.

In order to prove Theorem 2(b) I will show that

$$\limsup_{t\uparrow\infty} W(\xi,t) \le \alpha(\lambda) \tag{8}$$

and

$$\liminf_{t\uparrow\infty} W(\xi,t) \ge \alpha(\lambda) \tag{9}$$

where $\alpha = \alpha(\lambda)$ is the unique solution of

$$1 - \alpha = e^{-\alpha/\lambda}$$

in $(0,1)$. I will begin by proving (8) which is a consequence of the following analog of Proposition 2.

Proposition 3. Let W denote the solution of the initial value problem (5), where v_0 satisfies (2), k satisfies (K), and $\lambda \in (0,1)$. Then

$$W(\xi,t) \le \alpha + \frac{m(\mu)e^{-\mu\xi}}{|\mu E_\lambda(\mu)|} \mid \exp[\mu\{E_\lambda(\mu) - c\}t] - \exp[-c\mu t]|$$

in $R \times R^+$ for every $\mu \in (0,\nu)$ such that $E_\lambda(\mu) \neq 0$, where

$$m(\mu) \equiv \lambda \sup_R e^{\mu\xi} v_0(\xi)$$

and

$$E_\lambda(\mu) \equiv \frac{1}{\mu} [\{1 - \alpha(\lambda)\} \int_R e^{\mu\xi} k(\xi) d\xi - \lambda].$$

Proof. Set $W(\xi,t) \equiv \alpha + \Phi(\xi,t)$, where $\alpha = \alpha(\lambda)$. Then, since $e^{-\frac{\alpha}{\lambda}} = 1 - \alpha$, it follows that Φ satisfies

$$\Phi_t = c\Phi_\xi - \lambda(\alpha + \Phi) + \lambda\{1 - (1 - \alpha)e^{-\frac{1}{\lambda} k*\Phi}\} + \lambda v_0(\xi + ct) \quad \text{in } R \times R^+$$

$$\Phi(\xi,0) = -\alpha \quad \text{in } R.$$

Let Ψ denote the solution of the linear initial value problem

$$\Psi_t = c\Psi_\xi - \lambda\Psi + (1 - \alpha)k*\Psi + \lambda v_0(\xi + ct) \quad \text{in} \quad R \times R^+$$

$$\Psi(\xi,0) = 0 \quad \text{in} \quad R.$$

Then $\Psi(\xi,0) > \Phi(\xi,0)$ in R and

$$\Psi_t - c\Psi_\xi + \lambda(\alpha + \Psi) - \lambda\{1 - (1 - \alpha)e^{-\frac{1}{\lambda}k*\Psi}\}$$

$$\geq \Psi_t - c\Psi_\xi + \lambda\Psi - (1 - \alpha)k*\Psi$$

$$= \lambda v_0 = \Phi_t - c\Phi_\xi + \lambda(\alpha + \Phi) - \lambda\{1 - (1-\alpha)e^{-\frac{1}{\lambda}k*\Phi}\}$$

in $R \times R^+$. Thus, by the Comparison Lemma, $\Psi \geq \Phi$ in $R \times R^+$. The remainder of the proof is essentially the same as the proof of Proposition 2 and I will omit further details.

Proof of (8). In view of Proposition 3 it suffices to show that there exists a $\mu \in (0,\nu)$ such that $E_\lambda(\mu) \neq 0$ and $E_\lambda(\mu) - c < 0$. For $\lambda \in (0,1)$ define

$$F_\lambda(s) \equiv 1 - s - e^{-s/\lambda} .$$

Then $F_\lambda(0) = F_\lambda(\alpha(\lambda)) = 0$. Moreover, it is easy to verify that F_λ is strictly increasing for $s \in (0, -\lambda \log \lambda)$ and strictly decreasing for $s > -\lambda \log \lambda$ with $F_\lambda \downarrow -\infty$ as $s \uparrow \infty$. In particular, $\alpha(\lambda) > -\lambda \log \lambda$ which implies that

$$1 - \alpha(\lambda) = \exp\{-\alpha(\lambda)/\lambda\} < \lambda .$$

Therefore, since

$$\lim_{\mu\downarrow 0} [\{1 - \alpha(\lambda)\} \int_R e^{\mu\xi}k(\xi)d\xi - \lambda] = 1 - \alpha(\lambda) - \lambda < 0$$

it follows that $E_\lambda(\mu) \to -\infty$ as $\mu \downarrow 0$. Note that (8) holds for every $c > 0$, not just for $c \in (0,c^*)$.

The proof of (9) depends on the construction of a subsolution for problem (5). Specifically, I will construct a function $\psi(\xi)$ such that ψ has compact support in R, $\psi \in [0,\varepsilon]$ for some small $\varepsilon > 0$, and $Y(\xi,t) \uparrow \alpha(\lambda)$ as $t \uparrow \infty$ where Y is the solution of the initial value problem

$$Y_t = Q_c[Y] \quad \text{in} \quad R \times R^+$$

$$Y(\xi,0) = \psi(\xi) \quad \text{in} \quad R$$

The construction is based on the following rather technical lemmas. In stating them I will use the notation

$$L_{\eta c}[v] \equiv cv' - \lambda v + \eta(k*v) .$$

<u>Lemma 1</u>. For any $c \in (0,c^*)$ there exists $\gamma_1 \in R^+$, $\eta_1 \in (0,1)$ and

$$\mu_c : (0, \gamma_1) \times (\eta_1, 1) \to (0,\mu^*)$$

such that for $\gamma \in (0, \gamma_1)$ and $\eta \in (\eta_1, 1)$

$$L_{\eta c}[g_{\gamma\eta}](\xi) > 0 \quad \text{in} \quad (0, \pi/\gamma)$$

where

$$g_{\gamma\eta}(\xi) = e^{-\mu_c(\gamma,\eta)\xi} \sin \gamma\xi$$

<u>Proof</u>. Let $\gamma \in R^+$, $\eta \in (0,1)$, and $\mu \in (0,\nu)$ be arbitrary and set $v = e^{-\mu\xi} \sin \gamma\xi$. Then

$$L_{\eta c}[v](\xi) = \{-c\mu - \lambda + \eta \int_R e^{\mu\zeta}k(\zeta) \cos \gamma\zeta \, d\zeta\}v(\xi)$$
$$+ \{c\gamma - \eta \int_R e^{\mu\zeta}k(\zeta) \sin \gamma\zeta \, d\zeta\}e^{-\mu\xi} \cos \gamma\xi \quad .$$

Therefore $L_{\eta c}[v](\xi) > 0$ in $(0, \pi/\gamma)$ if γ, η, and μ are such that

$$c = \frac{\eta}{\gamma} \int_R k(\zeta)e^{\mu\zeta} \sin \gamma\zeta \, d\zeta \equiv \eta B(\gamma,\eta)$$

and

$$c < \frac{1}{\mu} \{\eta \int_R k(\zeta) e^{\mu\zeta} \cos \gamma\zeta \, d\zeta - \lambda\} \equiv A_\lambda(\gamma, \eta, \mu) .$$

Let

$$B(\mu) \equiv \int_R \zeta \, k(\zeta) e^{\mu\zeta} d\zeta .$$

Then $B(0) = 0$ and B is an increasing function of μ in $(0,\nu)$. Since

$$A_\lambda'(\mu) = \frac{1}{\mu} \{B(\mu) - A_\lambda(\mu)\}$$

it follows from (K3) that $B(\mu) < A_\lambda(\mu)$ in $(0,\mu^*)$ and $B(\mu^*) = A_\lambda(\mu^*)$. Fix $\eta_0 \in (0,1)$ such that $\eta_0 B(\mu^*) \equiv c_1 > c$. Now choose an arbitrary $c_0 \in (c, c_1)$ and define μ_0 by $\eta_0 B(\mu_0) = c_0$. Clearly $\mu_0 \in (0,\mu^*)$ and

$$\lim_{\gamma \downarrow 0} \mathcal{B}(\gamma,\mu) = B(\mu)$$

uniformly for $\mu \in [0, \mu_0]$. For arbitrary $\delta > 0$ which satisfies $\delta < c < c_0 - \delta$ and $B(\mu_0) < c^* - \delta$ there exists a $\gamma_0 = \gamma_0(\delta) > 0$ such that $\gamma \in (0, \gamma_0)$ implies

$$B(\mu) - \delta < \mathcal{B}(\gamma,\mu) < B(\mu) + \delta \quad \text{for all} \quad \mu \in [0, \mu_0].$$

In particular, for $\gamma \in (0, \gamma_0)$ and $\eta \in (\eta_0, 1)$

$$\eta\mathcal{B}(\gamma,\mu) < \eta\{B(\mu) + \delta\} < B(\mu) + \delta < c^* \quad \text{for} \quad \mu \in [0, \mu_0]$$

and

$$\eta\mathcal{B}(\gamma, \mu_0) > \eta B(\mu_0) - \delta\mu = \frac{\eta}{\eta_0} c_0 - \eta\delta > \frac{\eta}{\eta_0} (c_0 + \delta) - \eta\delta > c.$$

On the other hand, $\eta\mathcal{B}(\gamma,0) = 0$. Thus there exists

$$\mu_c : (0, \gamma_0) \times (\eta_0, 1) \to (0, \mu_0)$$

such that $\eta B(\gamma, \mu_c(\gamma,\eta)) = c$. Since

$$c = \eta B(\gamma, \mu_c(\gamma,\eta)) < B(\mu_c(\gamma,\eta)) + \delta$$

it follows that $\mu_c(\gamma,\eta) > B^{-1}(c - \delta) \equiv \bar{\mu}_c > 0$.

It is easily verified that

$$\lim_{\gamma\downarrow 0,\eta\uparrow 1} A_\lambda(\gamma, \eta, \mu) = A_\lambda(\mu)$$

uniformly in $[\bar{\mu}_c, \mu_0]$. Moreover, $A_\lambda(\mu) \geq A_\lambda(\mu_0) > c^*$ in $[\bar{\mu}_c, \mu_0]$. Thus there exist numbers $\gamma_1 \in (0, \gamma_0]$ and $\eta_1 \in [\eta_0, 1)$ such that $\gamma \in (0, \gamma_1)$ and $\eta \in (\eta_1, 1)$ imply $A_\lambda(\gamma, \eta, \mu) > c^* > c$ in $[\bar{\mu}_c, \mu_0]$. In particular, since $\mu_c(\gamma,\eta) \in [\bar{\mu}_c, \mu_0]$ it follows that $A_\lambda(\gamma, \eta, \mu_c(\gamma,\eta)) > c$ for $\gamma \in (0, \gamma_1)$ and $\eta \in (\eta_1, 1)$.

<u>Lemma 2</u>. Let

$$\psi_{\gamma\eta}(\xi) = \begin{cases} e^{-\mu_c(\gamma,\eta)\xi} \sin \gamma\xi & \text{for } \xi \in [0, \pi/\gamma] \\ 0 & \text{for } \xi \notin [0, \pi/\gamma] \end{cases}$$

There exists a $\gamma_2 \in (0, \gamma_1]$ such that $\gamma \in (0, \gamma_2)$ and $\eta \in (\eta_1, 1)$ imply

$$L_{\eta c}[\psi_{\gamma\eta}](\xi) > 0 \quad \text{for} \quad \xi \in (0, \pi/\gamma)$$

<u>Proof</u>. Since $\psi_{\gamma\eta} = g_{\gamma\eta}$ in $[0, \pi/\gamma]$ it follows that

$$L_{\eta c}[\psi_{\gamma\eta}](\xi) = L_{\eta c}[g_{\gamma\eta}](\xi) + \eta\{k*(\psi_{\gamma\eta} - g_{\gamma\eta})\}(\xi) \quad \text{for} \quad \xi \in [0, \pi/\gamma].$$

According to Lemma 1, $\gamma \in (0, \gamma_1)$ and $\eta \in (\eta_1, 1)$ imply $L_{\eta c}[g_{\gamma c}] > 0$ in $(0, \pi/\gamma)$. Hence it suffices to show that, for sufficiently small $\gamma > 0$,

$$I(\xi) \equiv \{k*(\psi_{\gamma\eta} - g_{\gamma\eta})\}(\xi) = \int_0^{\pi/\gamma} k(\xi - \zeta)e^{-\mu_c\zeta} \sin \gamma\zeta \, d\zeta$$

$$- \int_{-\infty}^{\infty} k(\xi - \zeta)e^{-\mu_c\zeta} \sin \gamma\zeta \, d\zeta \geq 0$$

for $\xi \in (0, \pi/\gamma)$. The second integral can be rewritten as

$$\int_0^{\pi/\gamma} \{ \sum_{n=-\infty}^{\infty} (-1)^n e^{-\mu_c n\pi/\gamma} k(\xi - \zeta - \frac{n\pi}{\gamma})\} e^{-\mu_c\zeta} \sin \gamma\zeta \, d\zeta \ .$$

Thus $I(\xi) \geq 0$ in $(0, \pi/\gamma)$ if

$$k(\xi) \geq \sum_{n=-\infty}^{\infty} (-1)^n e^{-\mu_c n\pi/\gamma} k(\xi - \frac{n\pi}{\gamma}) \quad \text{for} \quad \xi \in (-\frac{\pi}{\gamma}, \frac{\pi}{\gamma}]. \tag{10}$$

Recall that $\mu_c = \mu_c(\gamma,\eta) \in [\bar{\mu}_c, \mu_0]$ for $\gamma \in (0, \gamma_1)$ and $\eta \in (\eta_1, 1)$. By a tedious but straightforward computation one can show that (K4) implies the existence of a $\gamma_2 \in (0, \gamma_1]$ such that (10) holds for all $\eta \in (\eta_1, 1)$ and $\gamma \in (0, \gamma_2)$.

<u>Lemma 3</u>. There exists an $\eta_2 \in [\eta_1, 1)$ and $\kappa = \kappa_\lambda(\eta) \in R^+$ such that $\gamma \in (0, \gamma_2)$, $\eta \in (\eta_2, 1)$, and $\varepsilon \in (0,\kappa)$ implies $\varepsilon\psi_{\gamma\eta} \leq \alpha(\lambda)$ in R and

$$Q_c[\varepsilon\psi_{\gamma\eta}](\xi) \begin{cases} > 0 & \text{for } \xi \in (0, \pi/\gamma) \\ \geq 0 & \text{for } \xi \notin [0, \pi/\gamma]. \end{cases}$$

<u>Proof</u>. For each $\eta \in (0,1)$ there exists a unique positive $\theta = \theta(\eta)$ such that $1 - e^{-\theta} = \eta\theta$ and $1 - e^{-s} > \eta s$ for $0 < s < \theta$. Thus

$$E[\varepsilon\psi_{\gamma\eta}] \geq \varepsilon\eta k*\psi_{\gamma\eta}$$

provided that $\varepsilon k*\psi_{\gamma\eta} \leq \lambda\theta(\eta)$. Moreover, there exists an $\eta_2 \in [\eta_1, 1)$ such that $\lambda\theta(\eta) \leq \alpha(\lambda)$ for all $\eta \in (\eta_2, 1)$. Set $\kappa = \kappa_\lambda(\eta) = \lambda\theta(\eta)$. Then $\eta \in (\eta_2, 1)$ and $\varepsilon \in (0,\kappa)$ implies $\varepsilon\psi_{\gamma\eta} \leq \lambda\theta(\eta) \leq \alpha(\lambda)$ and the assertion follows from Lemma 2.

Proposition 3. Let $Y(\xi,t)$ denote the solution of the initial value problem

$$\begin{aligned} &Y_t = Q_c[Y] \quad \text{in} \quad \mathcal{R} \times \mathcal{R}^+ \\ &Y(\xi,0) = \varepsilon\psi_{\gamma\eta}(\xi) \quad \text{in} \quad \mathcal{R} \end{aligned} \tag{11}$$

with $\gamma \in (0, \gamma_2)$, $\eta \in (\eta_2, 1)$ and $\varepsilon \in (0,\kappa)$. Then $Y(\xi,t) \uparrow \alpha(\lambda)$ as $t \uparrow \infty$.

Proof. Let $U = Y$ and $V = \varepsilon\psi_{\gamma\eta}$. Then $U(\xi,0) = V(\xi,0)$ and, in view of Lemma 3,

$$U_t - Q_c[U] = 0 \geq -Q_c[\varepsilon\psi_{\gamma\eta}] = V_t - Q_c[V].$$

Thus, by the Comparison Lemma, $Y(\xi,t) \geq \varepsilon\psi_{\gamma\eta}(\xi) = Y(\xi,0)$ in $\mathcal{R} \times \mathcal{R}^+$. Now apply the Comparison Lemma to $U = Y(\xi, t + h)$ and $V = Y(\xi,t)$ for arbitrary fixed $h > 0$ to conclude that $Y(\xi, t + h) \geq Y(\xi,t)$ in $\mathcal{R} \times \mathcal{R}^+$. Thus Y is a nondecreasing function of t for each fixed $\xi \in \mathcal{R}$. Moreover, an application of the Comparison Lemma with $U \equiv \alpha(\lambda)$ and $V = Y$ shows that $Y \leq \alpha(\lambda)$ in $\mathcal{R} \times \mathcal{R}^+$. Therefore

$$\lim_{t\uparrow\infty} Y(\xi,t) \equiv q(\xi)$$

exists in $\mathcal{R}$.

By formal integration of equation (11)

$$Y(\xi,t) = \varepsilon\psi_{\gamma\eta}(\xi + ct)e^{-\lambda t} + \int_0^t e^{-\lambda(t-\tau)}E[Y](\xi + c(t - \tau),\tau)d\tau.$$

Define a new variable of integration σ by $\sigma = \xi + c(t - \tau)$. Then

$$\begin{aligned} Y(\xi,t) &= \varepsilon\psi_{\gamma\eta}(\xi + ct)e^{-\lambda t} \\ &+ \frac{1}{c}\int_\xi^\infty e^{\frac{\lambda}{c}(\xi-\sigma)} E[Y](\sigma, t - \frac{1}{c}(\sigma - \xi))\chi_{[\xi,\xi+ct)}(\sigma)d\sigma . \end{aligned}$$

Since $Y \uparrow q$ it follows from the Monotone Convergence Theorem that

$$q(\xi) = \frac{1}{c} \int_{\xi}^{\infty} e^{\frac{\lambda}{c}(\xi-\sigma)} E[q](\sigma)d\sigma \quad .$$

Therefore

$$q' = \frac{\lambda}{c} q - E[q] \ , \tag{12}$$

that is, q is a solution of the steady state equation corresponding to (11). In particular, q is continuous in R.

By Proposition 1, $Y > 0$ in $R \times R^+$. In view of Lemma 3, $Y_t(\xi,0) > 0$ in $(0, \pi/\gamma)$ so that $Y(\xi,t) > \varepsilon\psi_{\gamma\eta}(\xi)$ in $(0, \pi/\gamma)$. Thus, in particular,

$$q(\xi) > \varepsilon\psi_{\gamma\eta}(\xi) \quad \text{in} \quad [0, \pi/\gamma].$$

Since q is continuous, $q(\xi + h) > \varepsilon\psi_{\gamma\eta}(\xi)$ in $[0, \pi/\gamma]$ and $q(\xi + h) \geq \varepsilon\psi_{\gamma\eta}(\xi)$ in R for all $h \in R$ such that $|h|$ is sufficiently small. Using the Comparison Lemma, it follows that $q(\xi + h) \geq q(\xi)$ for all $h \in R$ with $|h|$ sufficiently small and this implies that $q(\xi) \equiv$ constant. The only constant solutions of (12) are $q \equiv 0$ and $q \equiv \alpha(\lambda)$. However, $q(\xi) \not\equiv 0$ since $q > \varepsilon\psi_{\gamma\eta} > 0$ in $(0, \pi/\gamma)$. Therefore $q(\xi) \equiv \alpha(\lambda)$.

<u>Proof of (9)</u>. Fix $\gamma \in (0, \gamma_2)$, $\eta \in (\eta_2, 1)$ and $T > 0$. By Proposition 1

$$\max\{W(\xi,T) : 0 \leq \xi \leq \pi/\gamma\} \equiv m > 0 \ .$$

Choose $\varepsilon \in (0,\kappa)$ such that $\varepsilon\psi_{\gamma\eta}(\xi) < m$ in $[0, \pi/\gamma]$ and let $Y(\xi,t)$ denote the solution of the corresponding initial value problem for equation (11). Then $Y(\xi,0) \leq W(\xi,T)$ and it follows from the Comparison Lemma that $Y(\xi, t - T) \leq W(\xi,t)$ in $R \times [T, +\infty)$. Finally, $Y \uparrow \alpha(\lambda)$ as $t \uparrow \infty$ implies that (9) holds.

Remark. There is no essential difficulty in extending Theorem 1, Theorem 2(a) and the estimate (8) to the more realistic case of an epidemic in R^2. The proof of the analog of (9) is somewhat more difficult but it can be done (cf. [3] and [12]).

This work has been supported by the National Science Foundation under the Grant MCS 75-05074.

REFERENCES.

1. D. G. Aronson, Topics in Nonlinear Diffusion, CBMS/NSF Lecture Notes, SIAM, to appear 1977.

2. D. G. Aronson and H. F. Weinberger, Nonlinear diffusions in population genetics, combustion, and nerve propagation, Partial Differential Equations and Related Topics (J. Goldstein, editor), Lecture Notes in Math., Vol. 446, Springer-Verlag, 1975.

3. D. G. Aronson and H. F. Weinberger, Multidimensional nonlinear diffusions arising in population genetics, Adv. in Math., to appear 1977.

4. C. Atkinson and G. E. H. Reuter, Deterministic epidemic waves, Math. Proc. Camb. Phil. Soc., 80 (1976), 315-330.

5. H. E. Daniels, The deterministic speed of a simple epidemic, Perspectives in Probability and Statistics, Academic Press, 1976.

6. R. A. Fisher, The wave of advance of advantageous genes, Ann. of Eugenics, 7 (1936), 355-369.

7. D. G. Kendall, Mathematical models of the spread of infections, Mathematics and Computer Science in Biology and Medicine, Medical Research Council, 1965.

8. W. O. Kermack and A. G. McKendrick, A contribution to the mathematical theory of epidemics, Proc. Roy. Soc., A, 115 (1927), 700.

9. A. Kolmogoroff, I. Petrovsky and N. Piscounoff, Étude de l'équation de la diffusion ..., Bull. U. État Moscou, 6 (1937), 1-25.

10. D. Mollison, The rate of spatial propagation of simple epidemics, Proc. 6th Berkeley Symp. Math. Statist. and Prob., Vol. 3, 1972.

11. D. Mollison, Possible velocities for a simple epidemic, Adv. Appl. Prob., 4 (1972), 233-257.

12. H. F. Weinberger, Asymptotic behavior of a model in population genetics, to appear.

Donald G. Aronson
School of Mathematics
University of Minneapolis
Minneapolis, Minnesota 55455

J R CANNON and R E EWING

Galerkin procedures for systems of parabolic partial differential equations related to the transmission of nerve impulses

ABSTRACT

A continuous and a predictor-corrector discrete Galerkin method are considered for approximating solutions of boundary and initial-value problems for a quasi-linear parabolic system of partial differential equations which is coupled to a non-linear system of ordinary differential equations. A priori bounds that reduce the estimation of error to problems in approximation theory are derived. The predictor-corrector methods yield linear algebraic equations for strongly non-linear problems.

1. INTRODUCTION

Problems in genetics, nerve impulse transmission, and chemical reactor theory [1.6] have generated systems of semi-linear parabolic partial differential equations and coupled systems of semi-linear parabolic partial differential equations and non-linear ordinary differential equations. It is clear that the next generation of models will involve some coupled systems of quasi-linear parabolic partial differential equations and non-linear ordinary differential equations. At some point the power of analytic techniques will wane and be replaced with massive computational programs to generate and catalog the stable and unstable solution regions in the data space as well as to observe the type of stable solution that is obtained. Indeed, such computation efforts are probably already underway.

Consider the degenerate system of non-linear initial-boundary value problems given by:

$$\frac{\partial U^i(x,t)}{\partial t} = \sum_{\ell=1}^{N_1} \frac{\partial}{\partial x_\ell} \left(\sum_{j=1}^{N_2} \sum_{m=1}^{N_1} C_{ij}^{\ell m} \frac{\partial U^j}{\partial x_m} + B_i^\ell \right) + F_i, \quad (x,t) \in Q_T,$$

(1.1)

$$\frac{\partial V^k(x,t)}{\partial t} = G_k \qquad , \quad (x,t) \in Q_T,$$

for $i = 1,2,\ldots,N_2$ and $k = 1,2,\ldots,N_3$ where x lies in a bounded domain $\Omega \subset R^{N_1}$, $0 < t \le T$, and the solution vectors $U = (U^1,U^2,\ldots,U^{N_2})$ and $V = (V^1,V^2,\ldots,V^{N_3})$ map $Q_T \equiv \Omega \times (0,T]$ into R^{N_2} and R^{N_3} respectively subject to the initial conditions

$$U^i(x,0) = U_0^i(x), \qquad x \in \Omega ,$$

(1.2)

$$V^k(x,0) = V_0^k(x), \qquad x \in \Omega .$$

The boundary $\partial\Omega$ of Ω and its inner-directed normal $\nu = (\nu_1,\nu_2,\ldots,\nu_{N_1})$ are assumed as smooth as may be needed, and (1.1) is considered subject to the boundary conditions

$$\sum_{\ell=1}^{N_1} \left(\sum_{j=1}^{N_2} \sum_{m=1}^{N_1} C_{ij}^{\ell m} \frac{\partial U^j}{\partial x_m} + B_i^\ell \right) \nu_\ell = g_i(x,t) \ , \quad \text{on} \quad S_T, \tag{1.3}$$

where $g = (g_1,g_2,\ldots,g_{N_2})$ is a given map of $S_T \equiv \partial\Omega \times (0,T]$ into R^{N_2}. We shall assume the following functional dependences for the appropriate indices: $C_{ij}^{\ell m} = C_{ij}^{\ell m}(x,t,U,V)$, $B_i^\ell = B_i^\ell(x,t,U,V)$, $F_i = F_i(x,t,U,V)$ and $G_k = G_k(x,t,U,V)$.

In this paper a continuous Galerkin method and a predictor-corrector discrete Crank-Nicolson-Galerkin method for (1.1)-(1.3) are formulated and analyzed. The methods and analysis presented are generalizations of those of Douglas and Dupont [4]. In section two, some notation, definitions and the variational form of (1.1)-(1.3) are noted. Section 3 is devoted to the derivation of the Galerkin methods. An a priori estimate of the error for the continuous Galerkin method is derived in section 4. The analysis of a priori error for the predictor-corrector discrete Crank-Nicolson-Galerkin method is carried out in section 5. The estimates reduce the error estimation to a problem in approximation theory. Section 6 concludes the paper with a discussion of generalizations of the error estimates when applied to modifications of (1.1)-(1.3) and the presentation of a predictor-corrector-corrector Crank-Nicolson-Galerkin method.

It is worth emphasizing that the predictor-corrector and the predictor-corrector-corrector Crank-Nicolson-Galerkin methods yield linear systems of equations even though the underlying problem is quite non-linear.

2. PRELIMINARIES

The formulation of (1.1)-(1.3) will be condensed through the combined use of summation convention on repeated indices, $\dot{U}$ for $\frac{\partial U}{\partial t}$, $U_{,m}$ for $\frac{\partial U}{\partial x_m}$, and suppression of the arguments. Thus the differential system (1.1) and boundary system (1.3) become:

$$
\begin{aligned}
&\text{a)} \quad \dot{U}^i = (C_{ij}^{\ell m} U^j_{,m} + B_i^\ell)_{,\ell} + F_i , && Q_T , \\
&\text{b)} \quad \dot{V}^k = G_k , && Q_T , \\
&\text{c)} \quad (C_{ij}^{\ell m} U^j_{,m} + B_i^\ell)\nu_\ell = g_i , && S_T .
\end{aligned}
\tag{2.1}
$$

The Euclidean norm of the appropriate dimension will be denoted $|u|^2 = u^i u^i$. We next define the inner product:

$$\langle a,b\rangle \equiv \int_\Omega a(x,t)b(x,t)dx . \tag{2.2}$$

Summing inner products of components we define:

$$<\eta,U>_{(1)} \equiv \int_\Omega \eta^i(x,t)U^i(x,t)dx \quad , \text{ sum on } i , \text{ and}$$
$$<\zeta,V>_{(2)} \equiv \int_\Omega \zeta^k(x,t)V^k(x,t)dx \quad , \text{ sum on } k , \tag{2.3}$$

with corresponding norms

$$||U(\cdot,t)||^2_{(1),\Omega} \equiv <U,U>_{(1)} \quad , \text{ and}$$
$$||V(\cdot,t)||^2_{(2),\Omega} \equiv <V,V>_{(2)} . \tag{2.4}$$

We next define the norms

$$||U||^2_{(1),Q_T} \equiv \int_0^T \int_\Omega U^i U^i \, dxdt \ , \text{ and}$$
$$||(U)_x||^2_{(1),Q_T} \equiv \int_0^T \int_\Omega U^i_{,m} U^i_{,m} \, dxdt \tag{2.5}$$

with similar norms indexed (2). We finally define the norm

$$||(U)_x||_{Q_T,\infty} = \sup_{(x,t)\in Q_T} \sup_{\substack{j=1,\dots,N_2 \\ i=1,\dots,N_1}} \left| \frac{\partial U^j(x,t)}{\partial x_i} \right| , \text{ and}$$
$$||U||_{(j),\Omega,\infty} = \sup_{t\in(0,T]} ||U||_{(j),\Omega} \quad , \ j = 1,2 . \tag{2.6}$$

Let $\{v^\alpha: \alpha = 1,2,\dots,N_4\} \subset H^1(\Omega)$ denote a basis of a subspace $M \subset H^1(\Omega)$. Let H_1 and H_2 be Hilbert spaces with respective norms given by

$$||\eta||^2_{H_1} \equiv ||\eta||^2_{(1),Q_T} + ||(\eta)_x||^2_{(1),Q_T} \quad \text{and}$$
$$||\zeta||^2_{H_2} \equiv ||\zeta||^2_{(2),Q_T} \quad . \tag{2.7}$$

Let $M_1 \equiv \{\eta: \eta^i = \eta^i_\alpha(t)v^\alpha(x), \ i = 1,2,\dots,N_2\} \subset H_1$ and $M_2 \equiv \{\zeta: \zeta^k = \zeta^k_\alpha(t)v^\alpha(x), \ k = 1,2,\dots,N_3\} \subset H_2$. Multiplying (2.4,a) by

$\eta \in M_1$ and (2.4,b) by $\zeta \in M_2$ and integrating by parts we obtain $i = 1,2,\ldots,N_2$ and $k = 1,2,\ldots,N_3$,

$$
\begin{aligned}
&\langle\eta^i,\dot U^i\rangle = A^i(t;U,V;U\ ,\eta^i) - \int_{\partial\Omega\times\{t\}} \eta^i g_i ds \quad , \quad \eta \in M_1 \ , \\
&\langle\zeta^k,\dot V^k\rangle = G^k(t;U,V,\zeta) \quad , \quad \zeta \in M_2 \ , \\
&\langle\eta^i(x,0),U^i(x,0)\rangle = \langle\eta^i(x,0),U_0^i\rangle \quad , \quad \eta(x,0) \in M \ , \\
&\langle\zeta^k(x,0),V^k(x,0)\rangle = \langle\zeta^k(x,0),V_0^k\rangle \quad , \quad \zeta(x,0) \in M \ ,
\end{aligned}
\tag{2.8}
$$

where here in (2.8) we have not employed the summation convention and where we define for $a,d,\ e \in H_1$ and $b,f \in H_2$,

$$
\begin{aligned}
&A^i(t;a,b;d,e) \equiv C^i(t;a,b;d,e) + B^i(t;a,b;e) + F^i(t;a,b;e), \\
&C^i(t;a,b;d,e) \equiv - \int_\Omega e^i_{,\ell}\ C^{\ell m}_{ij}\ (x,t,a,b)d^j_{,m}\ \ dx \quad , \\
&B^i(t;a,b;e) \equiv - \int_\Omega e^i_{,\ell}\ B^\ell_i\ (x,t,a,b)dx \quad , \\
&F^i(t;a,b;e) \equiv - \int_\Omega e^i\ F_i(x,t,a,b)dx \quad , \text{ and} \\
&G^k(t;a,b;f) \equiv \int_\Omega f^k\ G_k(x,t,a,b)dx
\end{aligned}
\tag{2.9}
$$

and here in (2.9) we have not summed on i or k. We shall then define additional notation A, C, B, F and G to denote

$$
A = \sum_{i=1}^{N_2} A^i, \quad C = \sum_{i=1}^{N_2} C^i, \quad B = \sum_{i=1}^{N_2} B^i, \quad F = \sum_{i=1}^{N_2} F^i \ \text{ and } \ G = \sum_{k=1}^{N_3} G^k \ .
$$

We shall now state our main assumptions. For appropriate indices, $C^{\ell m}_{ij}$ is a continuous function on $\overline{Q}_T \times R^{N_2} \times R^{N_3}$ such that

$$
\lambda|y|^2 \le y^i_\ell\ C^{\ell m}_{ij}\ y^j_m \le \mu|y|^2 \ , \qquad y \in H_1 \ , \tag{2.10}
$$

where λ and μ are positive constants. For appropriate indices, $C_{ij}^{\ell m}$, B_i^{ℓ}, F_i, G_k satisfy the following Lipschitz type behavior

$$\begin{aligned}
|C_{ij}^{\ell m}(x,t,a,b) - C_{ij}^{\ell m}(x,t,c,d)| &\le D \quad , \\
|B_i^{\ell}(x,t,a,b) - B_i^{\ell}(x,t,c,d)| &\le D \quad , \\
|F_i(x,t,a,b) - F_i(x,t,c,d)| &\le D \quad , \\
|G_k(x,t,a,b) - G_k(x,t,c,d)| &\le D \quad ,
\end{aligned} \tag{2.11}$$

where

$$D = K(|a-c|^2 + |b-d|^2)^{\frac{1}{2}} . \tag{2.12}$$

Also, it is no loss of generality to assume that the data functions are smooth and bounded and that solutions of (1.1)-(1.3) are classical with bounded derivatives of any reasonable order.

3. GALERKIN PROCEDURES

We first define a continuous time Galerkin approximation of the variational problem given in (2.8). We shall require that U and V lie in a finite-dimensional subspace of $H^1(\Omega)(= W_2^1(\Omega))$ for each t†. We shall approximate U and V of (2.8) by W and Y where

$$\begin{aligned}
W^i &= w_\alpha^i(t)v^\alpha(x) \quad , \qquad i = 1,\dots,N_2 \ , \quad \text{and} \\
Y^k &= y_\alpha^k(t)v^\alpha(x) \quad , \qquad k = 1,\dots,N_3 \ .
\end{aligned} \tag{3.1}$$

By replacing U and V in (2.8) by W and Y and noting the separation of variables effect of (3.1) we obtain the following system of ordinary differential equations for the coefficients $w_\alpha^i(t)$ and $y_\alpha^k(t)$, for

†See Remark 2 in Section 6.

$i = 1,2,\dots,N_2$, $k = 1,2,\dots,N_3$, $\alpha = 1,2,\dots,N_4$ and $\beta = 1,2,\dots,N_4$,

$$\begin{aligned}
&\text{a)} \quad \langle v^\alpha, v^\beta\rangle \dot{w}^i_\beta(t) = A^i(t;W,Y;W\ ,v^\alpha) - \int_{\partial\Omega\times\{t\}} v^\alpha g_i\,ds \ , \quad t \in (0,T], \\
&\text{b)} \quad \langle v^\alpha, v^\beta\rangle \dot{y}^k_\beta(t) = G^k(t;W,Y;v^\alpha) \ , \quad t \in (0,T], \\
&\text{c)} \quad \langle v^\alpha, v^\beta\rangle w^i_\beta(0) = \langle v^\alpha, U^i_o(x)\rangle \ , \\
&\text{d)} \quad \langle v^\alpha, v^\beta\rangle y^k_\beta(0) = \langle v^\alpha, V^k_o(x)\rangle \ ,
\end{aligned} \tag{3.2}$$

where here the repeated index β is summed. Clearly the solution of this coupled system of ordinary differential equations will provide the coefficients of the v^α as in (3.1) to yield the approximate solution W and Y of (2.8).

If we multiply (3.2,a) and (3.2,c) by $\eta^i_\alpha(t)$ and then multiply (3.2,b) and (3.2,d) by $\zeta^i_\alpha(t)$ and sum on α we obtain for each $i = 1,2,\dots,N_2$ and $k = 1,2,\dots,N_3$,

$$\begin{aligned}
&\langle \eta^i, \dot{W}^i\rangle = A^i(t;W,Y;W\ ,\eta) - \int_{\partial\Omega\times\{t\}} \eta^i g_i\,ds \ , \\
&\langle \zeta^k, \dot{Y}^k\rangle = G^k(t;W,Y;\zeta) \ , \\
&\langle \eta^i(x,0), W^i(x,0)\rangle = \langle \eta^i(x,0), U^i_0\rangle \ , \\
&\langle \zeta^k(x,0), Y^k(x,0)\rangle = \langle \zeta^k(x,0), V^k_0\rangle \ .
\end{aligned} \tag{3.3}$$

This is our continuous time Galerkin approximation to the variational problem.

We now consider an approximation of (3.3) in which the time variable is discretized. Let $t_n = n\,\Delta\,t$ where $\Delta t = T/N$ for some positive integer N. Let f_n denote $f(t_n)$. Let W_* and Y_* be approximations of W and Y, respectively, which satisfy

$$<\eta,\frac{W_{*n+1}-W_{*n}}{\Delta t}> = A(t_{n+\frac{1}{2}}\ ;\ \frac{W_{*n+1}+W_{*n}}{2},\ \frac{Y_{*n+1}+Y_{*n}}{2};\ \frac{W_{*n+1}+W_{*n}}{2},\ \eta)$$

$$-\int_{\partial\Omega\times\{t_{n+\frac{1}{2}}\}} \eta\ g(x,t_{n+\frac{1}{2}})ds, \qquad n \geq 0,$$

$$<\zeta,\frac{Y_{*n+1}-Y_{*n}}{\Delta t}> = G(t_{n+\frac{1}{2}};\ \frac{W_{*n+1}+W_{*n}}{2},\ \frac{Y_{*n+1}+Y_{*n}}{2};\ \zeta), \qquad n \geq 0,$$

$$(3.4)$$

$$<\eta,W_{*0}> = <\eta,U_0> ,$$

$$<\zeta,Y_{*0}> = <\zeta,V_0> ,$$

for all i, k and α where the notation has suppressed the dependence on i, k and α and for all η and ζ of the form

$$\eta^i = \eta^i_\alpha\ v^\alpha(x) \ , \qquad i = 1,2,\ldots,N_2 \ , \quad \text{and}$$

$$\zeta^k = \zeta^k_\alpha\ v^\alpha(x) \ , \qquad k = 1,2,\ldots,N_3 \ , \qquad (3.5)$$

where the v^α are the basis for M. We shall call (3.4) our Crank-Nicolson-Galerkin approximation. We note that since A and G are non-linear functions, the solution of the difference system (3.4) requires the solution of a non-linear algebraic system at each time step. In order to avoid the need to solve these non-linear systems we introduce the following predictor-corrector approximation to (3.4) :

$$\text{a)}\quad <\eta,\frac{W_{**n+1}-W_{*n}}{\Delta t}> = A(t_{n+\frac{1}{2}}\ ;\ W_{*n},\ Y_{*n}\ ;\ \frac{W_{**n+1}+W_{*n}}{2},\ \eta)$$

$$-\int_{\partial\Omega\times\{t_{n+\frac{1}{2}}\}} \eta g ds \qquad , \quad n \geq 0 \ ,$$

b) $<\zeta,\frac{Y_{**n+1}-Y_{*n}}{\Delta t}> = G(t_{n+\frac{1}{2}} ; \ W_{*n}, Y_{*n} ; \zeta) \quad , n \geq 0 ,$

c) $<\eta,\frac{W_{*n+1}-W_{*n}}{\Delta t}> = A(t_{n+\frac{1}{2}} ; \frac{W_{**n+1}+W_{*n}}{2}, \frac{Y_{**n+1}+Y_{*n}}{2}; \frac{W_{*n+1}+W_{*n}}{2} ; \eta)$

$$-\int_{\partial\Omega\times\{t_{n+\frac{1}{2}}\}} \eta g ds \qquad , n \geq 0 , \tag{3.6}$$

d) $<\zeta,\frac{Y_{*n+1}-Y_{*n}}{\Delta t}> = G(t_{n+\frac{1}{2}} ; \frac{W_{**n+1}+W_{*n}}{2}, \frac{Y_{**n+1}+Y_{*n}}{2}; \zeta) \ , \quad n \geq 0 ,$

e) $<\eta,W_{*0}> = <\eta,U_0>$

f) $<\zeta,Y_{*0}> = <\zeta,V_0>$

for all i, k, and α where again the notation has suppressed these dependencies and for all η and ζ which are of the form (3.5). The system (a), (b), (e) and (f) of (3.6) are solved as a predictor for the intermediate values W_{**n+1} and Y_{**n+1}. Then these values are used in the corrector system (c), (d), (e) and (f) of (3.6) to solve for W_{*n+1} and Y_{*n+1}. We note that the algebraic problem for (3.6) is linear since it requires the solution of two linear algebraic systems at each time step when it is written in a form analogous to (3.2).

4. A PRIORI ESTIMATE FOR THE CONTINUOUS TIME CASE

The error bounds derived in this section will be in terms of a norm of $U - \tilde{\eta}$ and $V - \tilde{\zeta}$ where U and V are solutions to (2.8) and $\tilde{\eta}$ and $\tilde{\zeta}$ are the "best possible" approximations to U and V at a particular time from the space M. The problem of obtaining error bounds is thus reduced to a problem in approximation theory.

Theorem 4.1: There exist constants K_1 and K_2 which depend on T, N_1, N_2, N_3, N_4, λ, μ, K, $||(U)_x||_{Q_T,\infty}$, and diam Ω such that, for U and V a solution to (2.8), W and Y, a solution to (3.3), and $\tilde{\eta}$ any function from M_1 and $\tilde{\zeta}$ any function from M_2

$$||U-W||^2_{(1),\Omega,\infty} + ||V-Y||^2_{(2),\Omega,\infty} \leq \tag{4.1}$$

$$K_1\{||U-\tilde{\eta}||^2_{(1),\Omega,\infty} + ||(U-\tilde{\eta})_x||^2_{(1),Q_T} + ||\frac{\partial}{\partial t}(U-\tilde{\eta})||^2_{(1),Q_T}$$

$$+ ||V-\tilde{\zeta}||^2_{(2),\Omega,\infty} + ||\frac{\partial}{\partial t}(V-\tilde{\zeta})||^2_{(2),Q_T}\}$$

and

$$||(U-W)_x||^2_{(1),Q_T} \leq \tag{4.2}$$

$$K_2\{||U-\tilde{\eta}||^2_{(1),\Omega,\infty} + ||(U-\tilde{\eta})_x||^2_{(1),Q_T} + ||\frac{\partial}{\partial t}(U-\tilde{\eta})||^2_{(1),Q_T}$$

$$+ ||V-\tilde{\zeta}||^2_{(2),\Omega,\infty} + ||\frac{\partial}{\partial t}(V-\tilde{\zeta})||^2_{(2),Q_T}\} .$$

Remark: In all of the analysis that follows, the K's will denote positive constants which depend upon only the quantities noted for K_1 and K_2 in the statement of Theorem 4.1.

Proof: Sum (2.8) and (3.3) on all indices and difference to obtain

$$\langle\eta,\dot{U}-\dot{W}\rangle_{(1)} + \langle\zeta,\dot{V}-\dot{Y}\rangle_{(2)} =$$

$$A(t;U,V;U,\eta) - A(t;W,Y;W,\eta) + G(t;U,V;\zeta) - G(t;W,Y;\zeta) =$$

$$C(t;W,Y;U-W,\eta) + \{C(t;U,V;U,\eta) - C(t;W,Y;U,\eta)\} \tag{4.3}$$

$$+ \{B(t;U,V;\eta) - B(t;W,Y;\eta)\} + \{F(t;U,V;\eta) - F(t;W,Y;\eta)\}$$

$$+ \{G(t;U,V;\zeta) - G(t;W,Y;\zeta)\}.$$

We shall integrate (4.3) from 0 to t on τ and then use the following test functions:

$$\eta \equiv (U-W) + (\tilde{\eta}-U) = \tilde{\eta} - W \in M_1 \ , \quad \text{and}$$
$$\zeta \equiv (V-Y) + (\tilde{\zeta}-V) = \tilde{\zeta} - Y \in M_2 \ . \tag{4.4}$$

We split the terms on the left of (4.3) using (4.4). Integrating the first term on the left by parts we obtain

$$\int_0^t \int_\Omega \eta^i(\dot{U}^i-\dot{W}^i)dxd\tau = \tag{4.5}$$
$$\int_0^t \int_\Omega (U^i-W^i)(\dot{U}^i-\dot{W}^i)dxd\tau + \int_0^t \int_\Omega (\tilde{\eta}^i-U^i)(\dot{U}^i-\dot{W}^i)dxd\tau$$
$$= \tfrac{1}{2}||(U^i-W^i)(\cdot,t)||_\Omega^2 - \tfrac{1}{2}||(U^i-W^i)(\cdot,0)||_\Omega^2 + \int_\Omega(\tilde{\eta}^i-U^i)(U^i-W^i)\Big|_0^t dx -$$
$$- \int_0^t \int_\Omega (\dot{\tilde{\eta}}^i-\dot{U}^i)(U^i-W^i)dxd\tau \ .$$

A similar result holds for the second term on the left side of (4.3). We also split the first term on the right side of (4.3) using (4.4). We see that

$$C^i(t;W,Y;U-W,\eta) = C^i(t;W,Y;U-W,U-W) + C^i(t;W,Y;U-W,\tilde{\eta}-U) \tag{4.6}$$

holds for $i = 1,2,\ldots,N_2$. Summing (4.6) on i, using the positive definiteness of $C_{ij}^{\ell m}$ from (2.10) and the first terms of the splittings described above we obtain from (4.3)

$$\tfrac{1}{2}\{||(U-W)(\cdot,t)||_{(1),\Omega}^2 + ||(V-Y)(\cdot,t)||_{(2),\Omega}^2\} + \lambda||(U-W)_x||_{(1),Q_T}^2$$
$$\leq \tfrac{1}{2}\{||(U-W)(\cdot,0)||_{(1),\Omega}^2 + ||(V-Y)(\cdot,0)||_{(2),\Omega}^2\} \tag{4.7}$$

$$+ \left[\int_\Omega (\tilde{\eta}-U)^i (U-W)^i \Big|_t^0 dx + \int_\Omega (\tilde{\zeta}-V)^k (V-Y)^k \Big|_t^0 dx \right. \qquad \text{(4.7 Cont'd.)}$$

$$+ \left. \int_0^t \int_\Omega (\dot{\tilde{\eta}}-\dot{U})^i (U-W)^i dx d\tau + \int_0^t \int_\Omega (\dot{\tilde{\zeta}}-\dot{V})^k (V-Y)^k dx d\tau \right]$$

$$+ \int_0^t C(t;W,Y;U-W,\tilde{\eta}-U)d\tau + \int_0^t \{C(t;U,V;U,\eta) - C(t;W,Y;U,\eta)$$

$$+ \mathcal{B}(t;U,V;\eta) - \mathcal{B}(t;W,Y;\eta) + \mathcal{F}(t;U,V;\eta) - \mathcal{F}(t;W,Y;\eta)$$

$$+ \mathcal{G}(t;U,V;\zeta) - \mathcal{G}(t;W,Y;\zeta)\}d\tau \quad .$$

We shall use the Schwarz inequality on the terms in $[\cdot]$ in (4.7) and then the trivial inequality

$$ab \le \varepsilon a^2 + (4\varepsilon)^{-1} b^2 \qquad (4.8)$$

to obtain the necessary bounds. For example,

$$\left| \int_\Omega (\tilde{\eta}-U)^i (U-W)^i \Big|_0^t dx \right| \le ||(U-\tilde{\eta})(\cdot,t)||_{(1),\Omega} ||(U-W)(\cdot,t)||_{(1),\Omega}$$

$$+ ||(U-\tilde{\eta})(\cdot,0)||_{(1),\Omega} ||(U-W)(\cdot,0)||_{(1),\Omega}$$

$$\le (4\varepsilon_1^{-1}) ||(U-\tilde{\eta})(\cdot,t)||^2_{(1),\Omega} + \varepsilon_1 ||(U-W)(\cdot,t)||^2_{(1),\Omega} \qquad (4.9)$$

$$+ \tfrac{1}{2} ||(U-\tilde{\eta})(\cdot,0)||^2_{(1),\Omega} + \tfrac{1}{2} ||(U-W)(\cdot,0)||^2_{(1),\Omega}$$

$$\le \varepsilon_1 ||(U-W)(\cdot,t)||^2_{(1),\Omega} + K_3 ||U-\tilde{\eta}||^2_{(1),\Omega,\infty} + \tfrac{1}{2} ||(U-W)(\cdot,0)||^2_{(1),\Omega}$$

and

$$\left| \int_0^t \int_\Omega (\dot{\tilde{\eta}}-\dot{U})^i (U-W)^i dx d\tau \right| \le \int_0^t \left|\left|\frac{\partial}{\partial t} (U-\tilde{\eta})\right|\right|_{(1),\Omega} ||U-W||_{(1),\Omega} d\tau$$

$$\le \int_0^t \left\{ \tfrac{1}{2} \left|\left|\frac{\partial}{\partial t} (U-\tilde{\eta})\right|\right|^2_{(1),\Omega} + \tfrac{1}{2} ||U-W||^2_{(1),\Omega} \right\} d\tau \qquad (4.10)$$

$$\leq \tfrac{1}{2}\|\frac{\partial}{\partial t}(U-\tilde{\eta})\|^2_{(1),Q_T} + \tfrac{1}{2}\int_0^t \|U-W\|^2_{(1),\Omega}d\tau \quad . \qquad \text{(4.10 Cont'd.)}$$

We use the above techniques and (2.10) to obtain the following bound:

$$\left|\int_0^t C(t;W,Y;U-W,\tilde{\eta}-U)d\tau\right| = \left|-\int_0^t\int_\Omega (\tilde{\eta}^i-U^i)_{,\ell}C^{\ell m}_{ij}(x,t,W,Y)(U^j-W^j)_{,m}dxd\tau\right|$$

$$\leq \mu\int_0^t \|(\tilde{\eta}-U)_x\|_{(1),\Omega}\|(U-W)_x\|_{(1),\Omega}d\tau \qquad (4.11)$$

$$\leq \mu\,\varepsilon_2\|(U-W)_x\|^2_{(1),Q_T} + K_4(4\varepsilon^{-1}\mu)\|(\tilde{\eta}-U)_x\|^2_{(1),Q_T} \quad .$$

We next use (2.6), (2.11) and the techniques used above to obtain for the first part of η from (4.4),

$$\left|\int_0^t \{C(t;U,V;U,U-W) - C(t;W,Y;U,U-W)\}d\tau\right|$$

$$= \left|-\int_0^t\int_\Omega (U^i-W^i)_{,\ell}\{C^{\ell m}_{ij}(x,t,U,V) - C^{\ell m}_{ij}(x,t,W,Y)\}U^j_{,m}dxd\tau\right|$$

$$\leq K\|(U)_x\|_{Q_T,\infty}\int_0^t\int_\Omega |(U-W)_x|(|U-W|^2 + |V-Y|^2)^{\frac{1}{2}}\,dxd\tau \qquad (4.12)$$

$$\leq K\|(U)_x\|_{Q_T,\infty}\int_0^t \|(U-W)_x\|_{(1),\Omega}\{\|U-W\|_{(1),\Omega} + \|V-Y\|_{(2),\Omega}\}d\tau$$

$$\leq 2K\|(U)_x\|_{Q_T,\infty}\,\varepsilon_2\|(U-W)_x\|^2_{(1),Q_T} + K\|(U)_x\|_{Q_T,\infty}(4\varepsilon_2)^{-1}\int_0^t\{\|U-W\|^2_{(1),\Omega} + \|V-Y\|^2_{(2),\Omega}\}d\tau \ ,$$

where here K denotes a positive constant depending upon K of (2.12), N_1 and N_2. Similar results hold for the difference in the terms involving $\mathcal{B}$ except that there is no $\|(U)_x\|_{Q_T,\infty}$ term in those bounds. In a similar way we see that for the other part of η,

$$\left|\int_0^t \{C(t;U,V;U,\tilde{\eta}-U) - C(t;W,Y;U,\tilde{\eta}-U)\}d\tau\right| \qquad (4.13)$$

$$\leq K\|(U)_x\|_{Q_T,\infty}\|(U-\tilde{\eta})_x\|^2_{(1),Q_T}$$

$$+ \frac{K}{2} ||(U)_x||_{Q_T,\infty} \int_0^t \{||U-W||^2_{(1),\Omega} + ||V-Y||^2_{(2),\Omega}\}d\tau \, . \qquad \text{(4.13 Cont'd.)}$$

where here K has the same dependence as K in (4.12). Similarly we note that for some K_5 and K_6 we have

$$\begin{aligned} &| \int_0^t \{F(t;U,V;\eta) - F(t;W,Y;\eta)\}d\tau| \\ &\le K_5||U-\tilde{\eta}||^2_{(1),\Omega,\infty} + K_6 \int_0^t \{||U-W||^2_{(1),\Omega} + ||V-Y||^2_{(2),\Omega}\}d\tau \, . \end{aligned} \tag{4.14}$$

A similar result holds for the ΔG term with the $||U-\tilde{\eta}||^2_{(1),\Omega,\infty}$ bound replaced by $||V-\tilde{\zeta}||^2_{(2),\Omega,\infty}$. At this point we combine all the results of (4.7) through (4.14) and their counterparts. All terms with ε-multipliers are moved to the left side of (4.7) under the assumption that ε_j, $j = 1,2$ are sufficiently small. We obtain

$$\begin{aligned} &(\tfrac{1}{2} - \varepsilon_1)\{||(U-W)(\cdot,t)||^2_{(1),\Omega} + ||(V-Y)(\cdot,t)||^2_{(2),\Omega}\} \\ &+ (\lambda - 2K_7[||(U)_x||_{Q_T,\infty} + 1 + \mu]\varepsilon_2)||(U-W)_x||^2_{(1),Q_T} \\ &\le ||(U-W)(\cdot,0)||^2_{(1),\Omega} + ||(V-Y)(\cdot,0)||^2_{(2),\Omega} \\ &+ K_8||U-\tilde{\eta}||^2_{(1),\Omega,\infty} + K_9||V-\tilde{\zeta}||^2_{(2),\Omega,\infty} \\ &+ \tfrac{1}{2}\{||\tfrac{\partial}{\partial t}(U-\tilde{\eta})||^2_{(1),Q_T} + ||\tfrac{\partial}{\partial t}(V-\tilde{\zeta})||^2_{(2),Q_T}\} \\ &+ K_{10}(||(U)_x||_{Q_T,\infty})||(U-\tilde{\eta})_x||^2_{(1),Q_T} \\ &+ K_{11}\int_0^t \{||(U-W)(\cdot,\tau)||^2_{(1),\Omega} + ||(V-Y)(\cdot,\tau)||^2_{(2),\Omega}\}d\tau \\ &\le K_{12}\{||U-\tilde{\eta}||^2_{(1),\Omega,\infty} + ||V-\tilde{\zeta}||^2_{(2),\Omega,\infty} \\ &+ ||\tfrac{\partial}{\partial t}(U-\tilde{\eta})||^2_{(1),Q_T} + ||\tfrac{\partial}{\partial t}(V-\tilde{\zeta})||^2_{(2),Q_T} + ||(U-\tilde{\eta})_x||^2_{(1),Q_T}\} \\ &+ K_{13}\int_0^t \{||(U-W)(\cdot,\tau)||^2_{(1),\Omega} + ||(V-Y)(\cdot,\tau)||^2_{(2),\Omega}\}d\tau \end{aligned} \tag{4.15}$$

where we have noted that

$$||(U-W)(\cdot,0)||^2_{(1),\Omega} \le ||(U-\tilde{\eta})(\cdot,0)||^2_{(1),\Omega} \le ||U-\tilde{\eta}||^2_{(1),\Omega,\infty}\,, \quad \text{and}$$

$$||(V-Y)(\cdot,0)||^2_{(2),\Omega} \le ||(V-\tilde{\zeta})(\cdot,0)||^2_{(2),\Omega} \le ||V-\tilde{\zeta}||^2_{(2),\Omega,\infty}\,. \tag{4.16}$$

We can now drop the non-negative second term on the left of (4.15), let $\varepsilon_1 = \frac{1}{4}$, and use Gronwall's inequality to obtain the bound

$$\begin{aligned} &||(U-W)(\cdot,t)||^2_{(1),\Omega} + ||(V-Y)(\cdot,t)||^2_{(2),\Omega} \\ &\le K_{14}e^{4K_{13}T}\{||U-\tilde{\eta}||^2_{(1),\Omega,\infty} + ||(U-\tilde{\eta})_x||^2_{(1),Q_T} + ||\tfrac{\partial}{\partial t}(U-\tilde{\eta})||^2_{(1),Q_T} \\ &\qquad + ||V-\tilde{\zeta}||^2_{(2),\Omega,\infty} + ||\tfrac{\partial}{\partial t}(V-\tilde{\zeta})||^2_{(2),Q_T}\}\,. \end{aligned} \tag{4.17}$$

Then, since the right hand side of (4.17) is independent of t, we obtain the bound (4.1). We can then drop the non-negative first term on the left of (4.15) and use the bound (4.1) in the last term on the right of (4.15) to obtain the bound (4.2).

5. A PRIORI ESTIMATE FOR THE PREDICTOR-CORRECTOR CRANK-NICOLSON-GALERKIN APPROXIMATION. In this section we shall derive a bound for the error induced by using the predictor-corrector Crank-Nicolson-Galerkin system (3.6) to approximate the solution of the variational problem (2.8). The bound and techniques are similar to those used in the proof of Theorem 4.1. However, in this case we shall require additional smoothness of U in the variable t since t-derivatives have been replaced by differences. We shall show that the method is correct to the order of $(\Delta t)^{3/2}$. We note in the next section that if the Neumann boundary condition is replaced by a Dirichlet boundary condition or if the coefficients $C^{\ell m}_{ij}$ and B^j_ℓ are independent of U and V we can slightly alter the proof to obtain bounds which are correct to the order of $(\Delta t)^2$ which is the usual Crank-Nicolson order estimate. We

also present a predictor-corrector-corrector scheme which is correct to the order of $(\Delta t)^2$.

In order to simplify our calculations, we shall define some notation. Let W_* and Y_* be the solution to (3.6) with U and V the solution to (2.8). Let W_{**} and Y_{**} be the solution to the predictor in (3.6). Define the following error functions.

$$\begin{aligned} Z_n &= Z(x,t_n) \equiv U_n - W_{*n} \quad , \\ Z_{*n} &= Z_*(x,t_n) \equiv U_n - W_{**n} \quad , \\ X_n &= X(x,t_n) = V_n - Y_{*n} \quad , \text{ and} \\ X_{*n} &= X_*(x,t_n) = V_n - Y_{**n} \quad . \end{aligned} \tag{5.1}$$

We now abuse the notation of (5.1) to define for a function f of (x,t) or (x,t_n),

$$f_{n+\frac{1}{2}} = \tfrac{1}{2}[f_{n+1} + f_n] = \tfrac{1}{2}[f(x,t_{n+1}) + f(x,t_n)] \ . \tag{5.2}$$

We emphasize that $f_{n+\frac{1}{2}}$ is <u>not</u> $f(x,t_{n+\frac{1}{2}})$. Finally we shall abuse the notation still further by defining special intermediate values as follows:

$$\begin{aligned} W_{**n+\frac{1}{2}} &= \tfrac{1}{2}[W_{**n+1} + W_{*n}] \quad , \\ Y_{**n+\frac{1}{2}} &= \tfrac{1}{2}[Y_{**n+1} + Y_{*n}] \quad , \\ \tilde{Z}_{*n+\frac{1}{2}} &= \tfrac{1}{2}[Z_{*n+1} + Z_n] \quad , \text{ and} \\ \tilde{X}_{*n+\frac{1}{2}} &= \tfrac{1}{2}[X_{*n+1} + X_n] \quad . \end{aligned} \tag{5.3}$$

<u>Theorem 5.1:</u> Let U and V be a solution of (2.8) for $0 \le t \le T$ and let W_* and Y_* be a solution of (3.6) for $0 \le t_n \le T$. Let Z_n, Z_{*n}, X_n and X_{*n} be as in (5.1). Suppose that $\dfrac{\partial^3 U^i}{\partial t^2 \partial x_\ell}$, $\dfrac{\partial^3 U^i}{\partial t^3}$ and all of the second derivatives of U^i for $\ell = 1,2,\ldots,N_1$ and $i = 1,2,\ldots,N_2$ are continuous and bounded by $\tilde{K}$. Then there exists constants K_{15}, K_{16}, $\delta > 0$, and

$\tau_o > 0$ which depend upon $T, N_1, N_2, N_3, N_4, \lambda, \mu, K, \tilde{K}, ||(U)_x||_{Q_T,\infty}$ and diam Ω such that for $\Delta t \le \tau_o$, $\tilde{\eta} \in M_1$, $\tilde{\zeta} \in M_2$ and $n = 0,1,\ldots,N$, we have

$$
\begin{aligned}
&||Z_N||^2_{(1),\Omega} + ||X_N||^2_{(2),\Omega} + \delta \sum_{n=0}^{N-1} ||(Z_{n+\frac{1}{2}})_x||^2_{(1),\Omega}\Delta t \\
&\le K_{15}\{\sum_{n=0}^{N-1} ||(U-\tilde{\eta})_{n+\frac{1}{2}}||^2_{(1),\Omega}\Delta t + \sum_{n=0}^{N-1} ||((U-\tilde{\eta})_{n+\frac{1}{2}})_x||^2_{(1),\Omega}\Delta t \\
&+ \sum_{n=0}^{N-1} ||(V-\tilde{\zeta})_{n+\frac{1}{2}}||^2_{(2),\Omega}\Delta t + \sum_{n=1}^{N-1} ||\frac{(U-\tilde{\eta})_{n+\frac{1}{2}} - (U-\tilde{\eta})_{n-\frac{1}{2}}}{\Delta t}||^2_{(1),\Omega}\Delta t \\
&+ \sum_{n=1}^{N-1} ||\frac{(V-\tilde{\zeta})_{n+\frac{1}{2}} - (V-\tilde{\zeta})_{n-\frac{1}{2}}}{\Delta t}||^2_{(2),\Omega}\Delta t + ||(U-\tilde{\eta})_o||^2_{(1),\Omega} + ||(U-\tilde{\eta})_{\frac{1}{2}}||^2_{(1),\Omega} \\
&+ ||(U-\tilde{\eta})_{N-\frac{1}{2}}||^2_{(1),\Omega} + ||(V-\tilde{\zeta})_o||^2_{(2),\Omega} + ||(V-\tilde{\zeta})_{\frac{1}{2}}||^2_{(2),\Omega} \\
&+ ||(V-\tilde{\zeta})_{N-\frac{1}{2}}||^2_{(2),\Omega}\} + K_{16}(\Delta t)^3 .
\end{aligned}
\tag{5.4}
$$

<u>Proof</u>: The proof has two parts. Letting the error in the prediction of U_{n+1} and V_{n+1} be Z_{*n+1} and X_{*n+1} as in (5.1), we first obtain an estimate of the form

$$
\begin{aligned}
&||Z_{*n+1}||^2_{(1),\Omega} + ||X_{*n+1}||^2_{(2),\Omega} \le K_{17}[||Z_n||^2_{(1),\Omega} + ||X_n||^2_{(2),\Omega} \\
&+ ||(U-\tilde{\eta})_{n+\frac{1}{2}}||^2_{(1),\Omega} + ||((U-\tilde{\eta})_{n+\frac{1}{2}})_x||^2_{(1),\Omega} + ||(V-\tilde{\zeta})_{n+\frac{1}{2}}||^2_{(2),\Omega}] \\
&+ K_{18}(\Delta t)^3 .
\end{aligned}
\tag{5.5}
$$

The second part consists of using this bound in the argument for the corrector to produce the bound given by (5.4).

First we write (2.8) in the form of the predictor for (3.6) with test functions $\eta_{(1)}$ and $\zeta_{(1)}$:

$$
<\eta_{(1)}, \frac{U_{n+1}-U_n}{\Delta t} + E_n^{(1)}>_{(1)} + <\zeta_{(1)}, \frac{V_{n+1}-V_n}{\Delta t} + E_n^{(2)}>_{(2)} = \tag{5.6}
$$

$$= A(t_{n+\frac{1}{2}};U_n + E_n^{(3)},V_n + E_n^{(4)};U_{n+\frac{1}{2}} + E_n^{(5)},\eta_{(1)})$$

$$-\int_{\partial\Omega\times\{t_{n+\frac{1}{2}}\}} \eta_{(1)}^i g_i(x,t_{n+\frac{1}{2}})ds + G(t_{n+\frac{1}{2}};U_n + E_n^{(6)},V_n + E_n^{(7)};\zeta_{(1)}),$$

and for the corrector for (3.6) with test functions $\eta_{(2)}$ and $\zeta_{(2)}$:

$$<\eta_{(2)},\frac{U_{n+1}-U_n}{\Delta t} + E_n^{(8)}>_{(1)} + <\zeta_{(2)},\frac{V_{n+1}-V_n}{\Delta t} + E_n^{(9)}>_{(2)} = \tag{5.7}$$

$$= A(t_{n+\frac{1}{2}};U_{n+\frac{1}{2}} + E_n^{(10)},V_{n+\frac{1}{2}} + E_n^{(11)};U_{n+\frac{1}{2}} + E_n^{(12)},\eta_{(2)})$$

$$-\int_{\partial\Omega\times\{t_{n+\frac{1}{2}}\}} \eta_{(2)}^i g_i(x,t_{n+\frac{1}{2}})ds + G(t_{n+\frac{1}{2}});U_{n+\frac{1}{2}} + E_n^{(13)},V_{n+\frac{1}{2}} + E_n^{(14)};\zeta_{(2)}),$$

where

$$E_n^{(1)},E_n^{(2)},E_n^{(5)},E_n^{(8)},E_n^{(9)},E_n^{(10)},E_n^{(11)},E_n^{(12)},E_n^{(13)},E_n^{(14)} = O((\Delta t)^2) \text{ , and}$$

$$E_n^{(3)},E_n^{(4)},E_n^{(6)},E_n^{(7)} = O((\Delta t)) \tag{5.8}$$

pointwise and in L^2 . Subtracting corresponding parts of (3.6) from (5.6) and using (5.1) and (5.3), we obtain error equations for the predictor

$$<\eta_{(1)},\frac{Z_{*n+1}-Z_n}{\Delta t} + E_n^{(1)}>_{(1)} + <\zeta_{(1)}\frac{X_{*n+1}-X_n}{\Delta t} + E_n^{(2)}>_{(2)} =$$

$$= C(t_{n+\frac{1}{2}};W_{*n},Y_{*n};\tilde{Z}_{*n+\frac{1}{2}} + E_n^{(5)},\eta_{(1)})$$

$$+ \{C(t_{n+\frac{1}{2}};U_n + E_n^{(3)},V_n + E_n^{(4)};U_{n+\frac{1}{2}} + E_n^{(5)},\eta_{(1)}) - \tag{5.9}$$

$$- C(t_{n+\frac{1}{2}};W_{*n},Y_{*n};U_{n+\frac{1}{2}} + E_n^{(5)},\eta_{(1)})\}$$

$$+ \{B(t_{n+\frac{1}{2}};U_n + E_n^{(3)},V_n + E_n^{(4)};\eta_{(1)}) - B(t_{n+\frac{1}{2}};W_{*n},Y_{*n};\eta_{(1)})\}$$

$$+ \{F(t_{n+\frac{1}{2}};U_n + E_n^{(3)},V_n + E_n^{(4)};\eta_{(1)}) - F(t_{n+\frac{1}{2}};W_{*n},Y_{*n};\eta_{(1)})\}$$

(5.9 Cont'd.)

$$+ \{G(t_{n+\frac{1}{2}};U_n + E_n^{(6)},V_n + E_n^{(7)};\zeta_{(1)}) - G(t_{n+\frac{1}{2}};W_{*n},Y_{*n};\zeta_{(1)})\} .$$

Equation (5.9) corresponds to (4.3) in the proof of Theorem 4.1. As in that proof we let

$$\eta_{(1)} = \frac{Z_{*n+1}+Z_n}{2} + (\tilde{\eta}_{n+\frac{1}{2}} - U_{n+\frac{1}{2}}) = \tilde{Z}_{*n+\frac{1}{2}} + (\tilde{\eta}-U)_{n+\frac{1}{2}} , \quad \text{and}$$

$$\zeta_{(1)} = \frac{X_{*n+1}+X_n}{2} + (\tilde{\zeta}_{n+\frac{1}{2}} - V_{n+\frac{1}{2}}) = \tilde{X}_{*n+\frac{1}{2}} + (\tilde{\zeta}-V)_{n+\frac{1}{2}} , \tag{5.10}$$

and we split the first three terms using (5.10) and use (2.1) to obtain

$$\frac{1}{2\Delta t}\{||Z_{*n+1}||^2_{(1),\Omega} - ||Z_n||^2_{(1),\Omega} + ||X_{*n+1}||^2_{(2),\Omega}$$

$$- ||X_n||^2_{(2),\Omega}\} + \lambda||(\tilde{Z}_{*n+\frac{1}{2}})_x||^2_{(1),\Omega} \tag{5.11}$$

$$\leq <\tilde{Z}_{*n+\frac{1}{2}},\frac{Z_{*n+1}-Z_n}{\Delta t}>_{(1)} + <\tilde{X}_{*n+\frac{1}{2}}\frac{X_{*n+1}-X_n}{\Delta t}>_{(2)} - C(t_{n+\frac{1}{2}};W_{*n},Y_{*n};\tilde{Z}_{*n+\frac{1}{2}},\tilde{Z}_{*n+\frac{1}{2}})$$

$$\leq <(U-\tilde{\eta})_{n+\frac{1}{2}},\frac{Z_{*n+1}-Z_n}{\Delta t}>_{(1)} + <\eta_{(1)},E_n^{(1)}>_{(1)} + <(V-\tilde{\zeta})_{n+\frac{1}{2}},\frac{X_{*n+1}-X_n}{\Delta t}>_{(2)} +$$

$$+ <\zeta_{(1)},E_n^{(2)}>_{(2)} +$$

$$+ C(t_{n+\frac{1}{2}};W_{*n},Y_{*n};\tilde{Z}_{*n+\frac{1}{2}},(\tilde{\eta}-U)_{n+\frac{1}{2}}) + C(t_{n+\frac{1}{2}};W_{*n},Y_{*n};E_n^{(5)},\eta_{(1)})$$

$$+ \{C(t_{n+\frac{1}{2}};U_n + E_n^{(3)},V_n + E_n^{(4)};U_{n+\frac{1}{2}} + E_n^{(5)},\eta_{(1)})$$

$$- C(t_{n+\frac{1}{2}};W_{*n},Y_{*n};U_{n+\frac{1}{2}} + E_n^{(5)},\eta_{(1)})\} + \{B(t_{n+\frac{1}{2}};U_n + E_n^{(3)},V_n + E_n^{(4)};\eta_{(1)})$$

$$- B(t_{n+\frac{1}{2}};W_{*n},Y_{*n};\eta_{(1)})\} + \{F(t_{n+\frac{1}{2}};U_n + E_n^{(3)},V_n + E_n^{(4)},\eta_{(1)})$$

$$- F(t_{n+\frac{1}{2}};W_{*n},Y_{*n};\eta_{(1)})\} + \{G(t_{n+\frac{1}{2}};U_n + E_n^{(6)},V_n + E_n^{(7)};\zeta_{(1)})$$

$$- G(t_{n+\frac{1}{2}};W_{*n},Y_{*n};\zeta_{(1)})\}.$$

We now make straightforward applications of Schwarz' inequality and (4.8) as in the proof of Theorem 4.1. We first split all terms involving $||Z_{*n+1}||^2_{(1),\Omega}$, $||X_{*n+1}||^2_{(2),\Omega}$ and $||(Z_{*n+\frac{1}{2}})_x||^2_{(1),\Omega}$ apart from the others with ε_3 and ε_4 multipliers and transport these terms to the left side of (5.11) by taking ε_3 and ε_4 sufficiently small. We then break all other products of norms apart using (4.8) with $\varepsilon = \frac{1}{2}$. As in the proof of Theorem 4.1, we obtain

$$\begin{aligned}
&\frac{1}{2\Delta t}\{||Z_{*n+1}||^2_{(1),\Omega} + ||X_{*n+1}||^2_{(2),\Omega}\} + \lambda||(\tilde{Z}_{*n+\frac{1}{2}})_x||^2_{(1),\Omega} \\
&\le \frac{K_{19}}{\Delta t}\{||Z_n||^2_{(1),\Omega} + ||X_n||^2_{(2),\Omega}\} + \frac{\varepsilon_3}{\Delta t}\{||Z_{*n+1}||^2_{(1),\Omega} + ||X_{*n+1}||^2_{(2),\Omega}\} \\
&+ K_{20}\{\sum_{k=1}^{7}||E_n^{(k)}||^2_{\Omega} + ||(E_n^{(5)})_x||^2_{\Omega}\} \qquad (5.12) \\
&+ K_{21}\{||(U-\tilde{\eta})_{n+\frac{1}{2}}||^2_{(1),\Omega} + ||(V-\tilde{\zeta})_{n+\frac{1}{2}}||^2_{(2),\Omega}\} \\
&+ \frac{\varepsilon_3}{2}\{||Z_{*n+1}||^2_{(1),\Omega} + ||X_{*n+1}||^2_{(2),\Omega}\} + 2\mu\,\varepsilon_4||(\tilde{Z}_{*n+\frac{1}{2}})_x||^2_{(1),\Omega} \\
&+ K_{22}\mu||((U-\tilde{\eta})_{n+\frac{1}{2}})_x||^2_{(1),\Omega} + \varepsilon_4||(U)_x||^2_{Q_T,\infty}K||(\tilde{Z}_{*n+\frac{1}{2}})_x||^2_{(1),\Omega} \\
&+ K_{23}||(U)_x||_{Q_T,\infty}\{||((U-\tilde{\eta})_{n+\frac{1}{2}})_x||^2_{(1),\Omega} + ||Z_n||^2_{(1),\Omega} + ||X_n||^2_{(2),\Omega}\} \\
&+ K\varepsilon_4||(\tilde{Z}_{*n+\frac{1}{2}})_x||^2_{(1),\Omega} + 2K\varepsilon_3||Z_{*n+1}||^2_{(1),\Omega} + 2K\varepsilon_3||X_{*n+1}||^2_{(2),\Omega} \\
&+ K_{24}\{||(U-\tilde{\eta})_{n+\frac{1}{2}}||^2_{(1),\Omega} + ||(V-\tilde{\zeta})_{n+\frac{1}{2}}||^2_{(2),\Omega} + ||Z_n||^2_{(1),\Omega} + ||X_n||^2_{(2),\Omega}\}.
\end{aligned}$$

Transporting ε-terms, multiplying by Δt, and grouping like terms, we obtain for ε_3 and ε_4 sufficiently small

$$\begin{aligned}
&\{\tfrac{1}{2} - [1 + (\tfrac{1}{2} + 2K)\Delta t]\varepsilon_3\}\{||Z_{*n+1}||^2_{(1),\Omega} + ||X_{*n+1}||^2_{(1),\Omega}\} \\
&+ \{\lambda - [2\mu + 2K||(U)_x||_{Q_T,\infty} + K]\varepsilon_4\}\Delta t||(\tilde{Z}_{*n+\frac{1}{2}})_x||^2_{(1),\Omega} \qquad (5.13) \\
&\le K_{25}\{||Z_n||^2_{(1),\Omega} + ||X_n||^2_{(2),\Omega} + ||(U-\tilde{\eta})_{n+\frac{1}{2}}||^2_{(1),\Omega} + ||(V-\tilde{\zeta})_{n+\frac{1}{2}}||^2_{(2),\Omega}
\end{aligned}$$

(5.13 Cont'd.)

$$+ ||((U-\tilde{\eta})_{n+\frac{1}{2}})_x||^2_{(1),\Omega}\} + K_{26}\Delta t\{\sum_{k=1}^{7}||E_n^{(k)}||^2_\Omega + ||(E_n^{(5)})_x||^2_\Omega\} \quad .$$

If we assume $\Delta t < 1$, choose $\varepsilon_3 = (6 + 8K)^{-1}$, multiply both sides of (5.13) by 4, drop the non-negative second term on the left, and note that each of the $E_n^{(k)}$ and $(E_n^{(5)})_x$ are $O(\Delta t)$, we obtain the estimate (5.5).

If we subtract corresponding parts of (3.6) from (5.7) and use (5.1) and (5.3), we obtain for the corrector the error equation

$$\begin{aligned}
&<\eta_{(2)},\frac{Z_{n+1}-Z_n}{\Delta t} + E_n^{(8)}>_{(1)} + <\zeta_{(2)},\frac{X_{n+1}-X_n}{\Delta t} + E_n^{(9)}>_{(2)} \\
&\qquad = \mathcal{C}(t_{n+\frac{1}{2}};W_{**n+\frac{1}{2}},Y_{**n+\frac{1}{2}};Z_{n+\frac{1}{2}} + E_n^{(12)},\eta_{(2)}) \\
&+ \{\mathcal{C}(t_{n+\frac{1}{2}};U_{n+\frac{1}{2}} + E_n^{(10)},V_{n+\frac{1}{2}} + E_n^{(11)};U_{n+\frac{1}{2}} + E_n^{(12)},\eta_{(2)}) \\
&\qquad - \mathcal{C}(t_{n+\frac{1}{2}};W_{**n+\frac{1}{2}},Y_{**n+\frac{1}{2}};U_{n+\frac{1}{2}} + E_n^{(12)},\eta_{(2)})\} \\
&+ \{\mathcal{B}(t_{n+\frac{1}{2}};U_{n+\frac{1}{2}} + E_n^{(10)},V_{n+\frac{1}{2}} + E_n^{(11)};\eta_{(2)}) \\
&\qquad - \mathcal{B}(t_{n+\frac{1}{2}};W_{**n+\frac{1}{2}},Y_{**n+\frac{1}{2}};\eta_{(2)})\} \\
&+ \{\mathcal{F}(t_{n+\frac{1}{2}};U_{n+\frac{1}{2}} + E_n^{(10)},V_{n+\frac{1}{2}} + E_n^{(11)};\eta_{(2)}) \\
&\qquad - \mathcal{F}(t_{n+\frac{1}{2}};W_{**n+\frac{1}{2}},Y_{**n+\frac{1}{2}};\eta_{(2)})\} \\
&+ \{\mathcal{G}(t_{n+\frac{1}{2}};U_{n+\frac{1}{2}} + E_n^{(13)},V_{n+\frac{1}{2}} + E_n^{(14)};\zeta_{(2)}) \\
&\qquad - \mathcal{G}(t_{n+\frac{1}{2}};W_{**n+\frac{1}{2}},Y_{**n+\frac{1}{2}};\zeta_{(2)})\} \quad .
\end{aligned} \tag{5.14}$$

We now choose

$$\begin{aligned}
\eta_{(2)} &= \frac{Z_{n+1}+Z_n}{2} + (\tilde{\eta}_{n+\frac{1}{2}}-U_{n+\frac{1}{2}}) = Z_{n+\frac{1}{2}} + (\tilde{\eta}-U)_{n+\frac{1}{2}} \text{ , and} \\
\zeta_{(2)} &= \frac{X_{n+1}+X_n}{2} + (\tilde{\zeta}_{n+\frac{1}{2}}-V_{n+\frac{1}{2}}) = X_{n+\frac{1}{2}} + (\tilde{\zeta}-V)_{n+\frac{1}{2}} \quad .
\end{aligned} \tag{5.15}$$

Just as in (5.11) except transporting only $||(Z_{n+\frac{1}{2}})_x||^2_{(1),\Omega}$ terms, we obtain

$$\frac{1}{2\Delta t}\{||Z_{n+1}||^2_{(1),\Omega} + ||X_{n+1}||^2_{(2),\Omega} - ||Z_n||^2_{(1),\Omega} - ||X_n||^2_{(2),\Omega}\}$$

$$+ \{\lambda - [2\mu + 2K||(U)_x||_{Q_T,\infty} + K]\ \varepsilon_4\}||(Z_{n+\frac{1}{2}})_x||^2_{(1),\Omega}$$

$$\le <(U-\tilde{\eta})_{n+\frac{1}{2}}, \frac{Z_{n+1}-Z_n}{\Delta t}>_{(1)} + <(V-\tilde{\zeta})_{n+\frac{1}{2}}, \frac{X_{n+1}-X_n}{\Delta t}>_{(2)} \tag{5.16}$$

$$+ K_{27}\{||(U-\tilde{\eta})_{n+\frac{1}{2}}||^2_{(1),\Omega} + ||((U-\tilde{\eta})_{n+\frac{1}{2}})_x||^2_{(1),\Omega} + ||(V-\tilde{\zeta})_{n+\frac{1}{2}}||^2_{(2),\Omega}\}$$

$$+ K_{28}\{\sum_{k=8}^{14}||E_n^{(k)}||^2_\Omega + ||(E_n^{(12)})_x||^2_\Omega\} + K_{29}\{||Z_{n+1}||^2_{(1),\Omega} + ||X_{n+1}||^2_{(2),\Omega}$$

$$+ ||Z_n||^2_{(1),\Omega} + ||X_n||^2_{(2),\Omega} + ||Z_{*n+1}||^2_{(1),\Omega} + ||X_{*n+1}||^2_{(2),\Omega}\}$$

$$\le <(U-\tilde{\eta})_{n+\frac{1}{2}}, \frac{Z_{n+1}-Z_n}{\Delta t}>_{(1)} + <(V-\tilde{\zeta})_{n+\frac{1}{2}}, \frac{X_{n+1}-X_n}{\Delta t}>_{(2)}$$

$$+ K_{30}\{||(U-\tilde{\eta})_{n+\frac{1}{2}}||^2_{(1),\Omega} + ||((U-\tilde{\eta})_{n+\frac{1}{2}})_x||^2_{(1),\Omega} + ||(V-\tilde{\zeta})_{n+\frac{1}{2}}||^2_{(2),\Omega}$$

$$+ ||Z_{n+1}||^2_{(1),\Omega} + ||X_{n+1}||^2_{(2),\Omega} + ||Z_n||^2_{(1),\Omega} + ||X_n||^2_{(2),\Omega}\} + K_{31}(\Delta t)^3,$$

where we have utilized (5.5) and (5.8) in the last inequality. Multiplying both sides of (5.16) by $2\Delta t$, letting $K_{32} = 2K_{29} + 1$ and rearranging terms we see that for some $\lambda_1 > 0$ we have the bound

$$(1-K_{32}\Delta t)\{||Z_{n+1}||^2_{(1),\Omega} + ||X_{n+1}||^2_{(2),\Omega}\} \tag{5.17}$$

$$- (1 + K_{32}\Delta t)\{||Z_n||^2_{(1),\Omega} + ||X_n||^2_{(2),\Omega}\} + \lambda_1\Delta t||(Z_{n+\frac{1}{2}})_x||^2_{(1),\Omega}$$

$$\le K_{32}\Delta t\{||(U-\tilde{\eta})_{n+\frac{1}{2}}||^2_{(1),\Omega} + ||((U-\tilde{\eta})_{n+\frac{1}{2}})_x||^2_{(1),\Omega} + ||(V-\tilde{\zeta})_{n+\frac{1}{2}}||^2_{(2),\Omega}\}$$

$$+ 2K_{33}(\Delta t)^4 + 2\Delta t\{<(U-\tilde{\eta})_{n+\frac{1}{2}}, \frac{Z_{n+1}-Z_n}{\Delta t}>_{(1)} + <(V-\tilde{\zeta})_{n+\frac{1}{2}}, \frac{X_{n+1}-X_n}{\Delta t}>_{(2)}\}$$

$$- \Delta t\{||Z_n||^2_{(1),\Omega} + ||X_n||^2_{(2),\Omega}\}\ .$$

We now use a technique found in [4]. Set $h(\Delta t) = (1 - K_{32}\Delta t)(1 + K_{32}\Delta t)^{-1}$. Note that for $0 \le n\Delta t \le T$, $(h(\Delta t))^n$ is bounded above by 1 and below by a positive number γ which is independent of Δt, for Δt sufficiently small Multiply (5.17) by $(h(\Delta t))^n(1 + K_{32}\Delta t)^{-1}$ to obtain

$$
\begin{aligned}
&(h(\Delta t))^{n+1}\{||Z_{n+1}||^2_{(1),\Omega} + ||X_{n+1}||^2_{(2),\Omega}\} \qquad (5.18)\\
&- (h(\Delta t))^n\{||Z_n||^2_{(1),\Omega} + ||X_n||^2_{(2),\Omega}\} + \lambda_1\gamma\Delta t||(Z_{n+\frac{1}{2}})_x||^2_{(1),\Omega}\\
&\le K_{34}\Delta t\{||(U-\tilde{\eta})_{n+\frac{1}{2}}||^2_{(1),\Omega} + ||((U-\tilde{\eta})_{n+\frac{1}{2}})_x||^2_{(1),\Omega} + ||(V-\tilde{\zeta})_{n+\frac{1}{2}}||^2_{(2),\Omega}\}\\
&+ K_{35}(\Delta t)^4 + K_*\Delta t\{<(h(\Delta t))^n(U-\tilde{\eta})_{n+\frac{1}{2}}, \frac{Z_{n+1}-Z_n}{\Delta t}>_{(1)}\\
&+ <(h(\Delta t))^n(V-\tilde{\zeta})_{n+\frac{1}{2}}, \frac{X_{n+1}-X_n}{\Delta t}>_{(2)}\} - \gamma\Delta t\{||Z_n||^2_{(1),\Omega} + ||X_n||^2_{(2),\Omega}\},
\end{aligned}
$$

where $K_* = 2(1 + K_{32}\Delta t)^{-1}$. Next we sum on n from $n = 0,\ldots,N-1$ to obtain

$$
\begin{aligned}
&(h(\Delta t))^N\{||Z_N||^2_{(1),\Omega} + ||X_N||^2_{(2),\Omega}\} - \{||Z_o||^2_{(1),\Omega} + ||X_o||^2_{(2),\Omega}\}\\
&+ \sum_{n=0}^{N-1} \lambda_1\gamma\Delta t||(Z_{n+\frac{1}{2}})_x||^2_{(1),\Omega} \qquad (5.19)\\
&\le K_{36}\Delta t \sum_{n=0}^{N-1} \{||(U-\tilde{\eta})_{n+\frac{1}{2}}||^2_{(1),\Omega} + ||((U-\tilde{\eta})_{n+\frac{1}{2}})_x||^2_{(1),\Omega}\\
&+ ||(V-\tilde{\zeta})_{n+\frac{1}{2}}||^2_{(2),\Omega}\} + K_{37} \sum_{n=0}^{N-1} (\Delta t)^4 - \gamma\Delta t \sum_{n=0}^{N-1} \{||Z_n||^2_{(1),\Omega} + ||X_n||^2_{(2),\Omega}\}\\
&+ K_*\Delta t \sum_{n=0}^{N-1} \{<(h(\Delta t))^n(U-\tilde{\eta})_{n+\frac{1}{2}}, \frac{Z_{n+1}-Z_n}{\Delta t}>_{(1)}\\
&+ <(h(\Delta t))^n(V-\tilde{\zeta})_{n+\frac{1}{2}}, \frac{X_{n+1}-X_n}{\Delta t}>_{(2)}\}.
\end{aligned}
$$

We next sum the inner product terms by parts. For example, we obtain

$$
K_*\Delta t \sum_{n=0}^{N-1} <(h(\Delta t))^n(U-\tilde{\eta})_{n+\frac{1}{2}}, \frac{Z_{n+1}-Z_n}{\Delta t}>_{(1)} \qquad (5.20)
$$

(5.20 Cont'd.)

$$= -K_*\Delta t \sum_{n=0}^{N-1} < \frac{(h(\Delta t))^{n+1}(U-\tilde{\eta})_{n+1+\frac{1}{2}} - (h(\Delta t))^{n}(U-\tilde{\eta})_{n+\frac{1}{2}}}{\Delta t}, Z_{n+1}>(1)$$

$$+ K_*\Delta t<(h(\Delta t))^{N}(U-\tilde{\eta})_{N+\frac{1}{2}},Z_N>(1) - K_{33}\Delta t<(U-\tilde{\eta})_{\frac{1}{2}},Z_o>(1)$$

$$= -K_*\Delta t \sum_{n=1}^{N-1} < \frac{(h(\Delta t))^{n}(U-\tilde{\eta})_{n+\frac{1}{2}} - h(\Delta t)^{n-1}(U-\tilde{\eta})_{n-\frac{1}{2}}}{\Delta t}, Z_{n}>(1)$$

$$+ K_*\Delta t<(h(\Delta t))^{N-1}(U-\tilde{\eta})_{N-\frac{1}{2}},Z_N>(1) - K_{33}\Delta t<(U-\tilde{\eta})_{\frac{1}{2}}, Z_o>(1)$$

where the top term of the sum has been split off to cancel appropriately. As before we use Schwarz inequality and (4.8) to split (5.20) and its counterpart for V into sums of squares of the norms. We group the terms $\Delta t \sum_{n=1}^{N-1} ||Z_n||^2_{(1),\Omega}$ with an ε_5-multiplier and cancel it with the term with the γ-multiplier on the right side of (5.19) thus making the right side larger when $\varepsilon_5 < \gamma$. We take the $\Delta t||Z_N||^2_{(1),\Omega}$ term and transport it to the left side of (5.19) via an ε_6 sufficiently small. The result of the above manipulations is

$$\{\gamma - K_*\Delta t\varepsilon_6\}\{||Z_N||^2_{(1),\Omega} + ||X_N||^2_{(2),\Omega}\} + \sum_{n=0}^{N-1} \lambda_1\gamma\Delta t||(Z_{n+\frac{1}{2}})_x||^2_{(1),\Omega}$$

$$\leq K_{38}\Delta t \sum_{n=0}^{N-1}\{||(U-\tilde{\eta})_{n+\frac{1}{2}}||^2_{(1),\Omega} + ||((U-\tilde{\eta})_{n+\frac{1}{2}})_x||^2_{(1),\Omega} + ||(V-\tilde{\zeta})_{n+\frac{1}{2}}||^2_{(2),\Omega}\}$$

$$+ K_{39}\Delta t \sum_{n=1}^{N-1} \{||\frac{(U-\tilde{\eta})_{n+\frac{1}{2}}-(U-\tilde{\eta})_{n-\frac{1}{2}}}{\Delta t}||^2_{(1),\Omega} + ||\frac{(V-\tilde{\zeta})_{n+\frac{1}{2}}-(V-\tilde{\zeta})_{n-\frac{1}{2}}}{\Delta t}||^2_{(2),\Omega}\}$$

$$+ K_{40}\{||(U-\tilde{\eta})_{\frac{1}{2}}||^2_{(1),\Omega} + ||(U-\tilde{\eta})_{N-\frac{1}{2}}||^2_{(1),\Omega} + ||(V-\tilde{\zeta})_{\frac{1}{2}}||^2_{(2),\Omega}$$

$$+ ||(V-\tilde{\zeta})_{N-\frac{1}{2}}||^2_{(2),\Omega} + ||Z_o||^2_{(1),\Omega} + ||X_o||^2_{(2),\Omega}\} + K_{41}(\Delta t)^3 . \tag{5.21}$$

Finally, noticing $\Delta t < 1$, taking $\varepsilon_6 = [\gamma(2K_*)^{-1}]$, multiplying by $2\gamma^{-1}$, setting $\delta = 2\lambda_1$, increasing all constants appropriately and noting that

$$||Z_0||^2_{(1),\Omega} \le ||(U-\tilde{\eta})_0||^2_{(1),\Omega}, \quad \text{and}$$

$$||X_0||^2_{(2),\Omega} \le ||(V-\tilde{\zeta})_0||^2_{(2),\Omega}, \tag{5.22}$$

we obtain the desired bound (5.4) .

6. GENERALIZATIONS AND COMMENTS.

As presented in Theorem 5.1 the rate of convergence for the predictor-corrector Crank-Nicolson-Galerkin procedure was shown to be $O((\Delta t)^{3/2})$. The aim of the scheme was to achieve $O((\Delta t)^2)$. The problem lies in the coefficients on the right side of (5.11). Specifically, the terms of the predictor that are involved are

$$\begin{aligned}\{C(t_{n+\frac{1}{2}};U_n + E_n^{(3)},V_n + E_n^{(4)};U_{n+\frac{1}{2}} + E_n^{(5)},\eta_{(1)})\\ - C(t_{n+\frac{1}{2}};W_{*n},Y_{*n};U_{n+\frac{1}{2}} + E_n^{(5)},\eta_{(1)})\}\end{aligned} \tag{6.1}$$

and

$$\{B(t_{n+\frac{1}{2}};U_n + E_n^{(3)},V_n + E_n^{(4)};\eta_{(1)}) - B(t_{n+\frac{1}{2}};W_{*n},Y_{*n};\eta_{(1)})\}. \tag{6.2}$$

Both of these terms contain spatial derivatives of $\eta_{(1)} = \tilde{Z}_{n+\frac{1}{2}} + (\tilde{\eta}-U)_{n+\frac{1}{2}}$. The spatial derivatives of the $\tilde{Z}_{n+\frac{1}{2}}$ term would have to be transported to the left side of (5.11) and incorporated into the $\lambda||(Z_{*n+\frac{1}{2}})_x||^2_{(1),\Omega}$ term with some small coefficient. But, the errors $E_n^{(3)}$ and $E_n^{(4)}$ are of order $O(\Delta t)$ and in order to achieve the desired accuracy of $O((\Delta t)^2)$ there is a need to boost these terms by a factor of (Δt). Herein lies the problem. Splitting with the usual $ab \le \varepsilon a^2 + K\varepsilon^{-1}b^2$ inequality with $\varepsilon = \Delta t$ and $a = E_n^{(3)}$ or $E_n^{(4)}$ would produce the correct boost in Δt but it would place Δt^{-1} in front of the spatial derivatives of $\tilde{Z}_{*n+\frac{1}{2}}$. The factor $(\Delta t)^{-1}$ can hardly be regarded as small. Recalling the substitution of the predictor into the corrector it can be seen that if the spatial derivatives could be moved off of $\tilde{Z}_{*n+\frac{1}{2}}$, then a splitting which placed a factor of $(\Delta t)^{-1}$ in front of $\tilde{Z}_{*n+\frac{1}{2}}$ would cause no problem in the subsequent analysis and an accuracy of $O((\Delta t)^2)$ would be obtained. Recognizing all of this, Douglas and Dupont [4] in considering a similar problem with zero Dirichlet boundary data carried out an integration by parts which removed the spatial derivatives from the $\tilde{Z}_{*n+\frac{1}{2}}$ term. Since we can make the same kind of calculations for our problem (1.1)-(1.3) with (1.3) replaced by zero Dirichlet boundary data, we can state our first generalization.

Theorem 6.1: For (1.1)-(1.3) with (1.3) replaced by zero Dirichlet boundary data; i.e., $U = 0$ on S_T, the error for the predictor-corrector Crank-Nicolson-Galerkin procedure is of the form of (5.4) with the $(\Delta t)^3$ term replaced by $(\Delta t)^4$.

A second generalization resulting from the discussion above is the obvious case of (1.1)-(1.3) in which the $C_{ij}^{\ell m}$ and B_i^{ℓ} do not depend upon U and V.

Theorem 6.2: For (1.1)-(1.3) in which the $C_{ij}^{\ell m}$ and B_i^{ℓ} do not depend upon U and V, the error for the predictor-corrector Crank-Nicolson-Galerkin procedure is of the form of (5.4) with the $(\Delta t)^3$ term replaced by $(\Delta t)^4$.

Remark 1: It is worth pointing out that the case of (1.1)-(1.3) with $C_{ij}^{\ell m}$ and B_i^{ℓ} independent of U and V is currently being employed in many physical models.

For problem (1.1)-(1.3) the integration by parts yields a term on the boundary which is $O(\Delta t)$. The employment of various trace estimates will not improve this situation. What is needed is a modification of the boundary term. Such a modification and its analysis are beyond the scope of this presentation and will be deferred until later [2, 3]. Also, see Douglas and Dupont [5].

Remark 2: By using a predictor-corrector-corrector scheme to replace (3.6) and thus solving three linear algebraic systems at each time step, we can obtain the desired $O((\Delta t)^2)$ Crank-Nicolson convergence in time for the original Neumann problem (1.1)-(1.3). We shall replace the corrector equations, (c) and (d) of (3.6), by the following corrector-corrector scheme:

$$
\begin{aligned}
&\text{(a)}\\
&\left\langle \eta, \frac{W_{***n+1}-W_{*n}}{\Delta t} \right\rangle = A\left(t_{n+\frac{1}{2}}; \frac{W_{**n+1}+W_{*n}}{2}, \frac{Y_{**n+1}+Y_{*n}}{2}; \frac{W_{***n+1}+W_{*n}}{2}, \eta\right)\\
&\qquad - \int_{\partial\Omega\times\{t_{n+\frac{1}{2}}\}} \eta g\, ds, \quad \eta \in M_1, \quad n \geq 0,\\
&\text{(b)}\\
&\left\langle \zeta, \frac{Y_{***n+1}-Y_{*n}}{\Delta t} \right\rangle = G\left(t_{n+\frac{1}{2}}; \frac{W_{**n+1}+W_{*n}}{2}, \frac{Y_{**n+1}+Y_{*n}}{2}; \zeta\right), \quad \zeta \in M_2, \quad n \geq 0,\\
&\text{(c)}\\
&\left\langle \eta, \frac{W_{*n+1}-W_{*n}}{t} \right\rangle = A\left(t_{n+\frac{1}{2}}; \frac{W_{***n+1}+W_{*n}}{2}, \frac{Y_{***n+1}+Y_{*n}}{2}; \frac{W_{*n+1}+W_{*n}}{2}, \eta\right)\\
&\qquad - \int_{\partial\Omega\times\{t_{n+\frac{1}{2}}\}} \eta g\, ds, \quad \eta \in M_1, \quad n \geq 0,\\
&\text{(d)}\\
&\left\langle \zeta, \frac{Y_{*n+1}-Y_{*n}}{\Delta t} \right\rangle = G\left(t_{n+\frac{1}{2}}; \frac{W_{***n+1}+W_{*n}}{2}, \frac{Y_{***n+1}+Y_{*n}}{2}; \zeta\right), \quad \zeta \in M_2, \quad n \geq 0.
\end{aligned}
\tag{6.3}
$$

We note that the major use of the corrector, (c) and (d) of (3.6), was to multiply the estimate from (5.5) by Δt and use the result in going from (5.16) to (5.17). The effect was to add another power of Δt to the last term in (5.5), which, after summing on n, became $K_{41}(\Delta t)^3$, the last term in (5.21). This term yielded the ultimate $O((\Delta t)^{3/2})$ time discretization error of the method. By consulting the analysis leading to (5.17) one can see that if (6.3)(a) and (b) were used in place of (3.6)(c) and (d), we could obtain an estimate of the form

$$||Z_{**n+1}||^2_{(1),\Omega} + ||X_{**n+1}||^2_{(2),\Omega} \le K_{42}[||Z_n||^2_{(1),\Omega} + ||X_n||^2_{(2),\Omega}$$
$$+ \Delta t\{||(U-\tilde{\eta})_{n+\frac{1}{2}}||^2_{(1),\Omega} + ||((U-\tilde{\eta})_{n+\frac{1}{2}})_x||^2_{(1),\Omega} \tag{6.4}$$
$$+ ||(V-\tilde{\zeta})_{n+\frac{1}{2}}||^2_{(2),\Omega}\}] + K_{43}(\Delta t)^4 ,$$

where

$$Z_{**n+1} = U_{n+1} - W_{***n+1}$$

and

$$X_{**n+1} = V_{n+1} - Y_{***n+1} \quad . \tag{6.5}$$

Then, just as (5.5) was used in (5.16), (6.4) can be used in the estimate for the second corrector equations (6.4)(c) and (d) to obtain a time discretization error of the form $O((\Delta t)^2)$ as desired. We have thus outlined a proof of the following result.

Theorem 6.3: If the predictor-corrector scheme (3.6) is replaced by a predictor-corrector-corrector scheme described by (3.6)(a) and (b), (6.3), and (3.6)(e) and (f), the error for this new approximation of the original problem (1.1)-(1.3) is of the form (5.4) with the $(\Delta t)^3$ term replaced by $(\Delta t)^4$.

Remark 3: Since spatial derivatives of V do not appear anywhere in the analysis above, another generalization of all of the above results is that of employing a subspace from $L^2(\Omega)$ which is distinct from $M \subset H^1(\Omega)$ for the approximation of Y of V.

REFERENCES.

1. D. G. Aronson, Topics in Non-linear Diffusion, to appear in the CBMS Regional Conference Series in Applied Mathematics.

2. J. R. Cannon and Richard E. Ewing, Galerkin Extrapolation Procedures for Mixed Systems of Quasi-Linear Parabolic Partial Differential Equations and Ordinary Differential Equations, to appear.

3. J. R. Cannon and Richard E. Ewing, Quasilinear Parabolic Systems with Nonlinear Boundary Conditions, to appear.

4. Jim Douglas, Jr. and T. Dupont, Galerkin Methods for Parabolic Equations, SIAM J. Numer. Anal., V. 7 (1970), pp. 575-626.

5. Jim Douglas, Jr. and T. Dupont, Galerkin Methods for Parabolic Equations with Non-linear Boundary Conditions, Numer. Math., V. 20 (1973), pp. 213-237.

6. S. P. Hastings, Some Mathematical Problems from Neurobiology, Amer. Math. Monthly, V. 82 (1975), pp. 881-894.

J. R. Cannon
Department of Mathematics
The University of Texas at Austin
Austin, Texas 78712

Richard E. Ewing
Department of Mathematics
University of Chicago
Chicago, Illinois 60637

E D CONWAY and J A SMOLLER*

Diffusion and the classical ecological interactions: asymptotics

1. INTRODUCTION

The Kolmogorov form of the ordinary differential equations describing the classical two species interactive growth is

$$\frac{dU}{dt} = UM(U,V)$$
$$\frac{dV}{dt} = VN(U,V) . \qquad (1)$$

The variables U , V denote certain measures of the total population such as number of individuals, mass, area of shade cast by plants, etc. [10], [11]. In this paper however, we shall assume that the quantities U and V are continuously distributed throughout a spatial domain, Ω , and shall concentrate our attention on spatial densities, u and v, related to U and V by

$$U(t) = \frac{1}{|\Omega|} \int_{\Omega} u(t,x)dx \ , \ V(t) = \frac{1}{|\Omega|} \int_{\Omega} v(t,x)dx \ .$$

The reason for normalizing by $|\Omega|$, the measure of Ω , will become clear in Section 4. If we further assume that the two populations are undergoing simple diffusion we are lead to study the system of partial differential equations

$$\begin{aligned} u_t &= d_1 \, \Delta u + uM(u,v) \\ v_t &= d_2 \, \Delta v + vN(u,v) \end{aligned} \qquad x \in \Omega \ , \ t > 0 \ , \qquad (2)$$

*Conway's research was supported in part by NSF grant MPS-73-08442 A02; Smoller's by AFOSR Contract No. AFOSR 71-2122 where d_1 , d_2 are non-negative constants. In a given situation the conditions imposed upon M and N in (2) might differ from the conditions that might be appropriate in the case of equation (1). In fact many of the usual heuristic biological arguments leading to assumptions defining a particular model seem much more compelling when M and N describe a local interaction than when they describe the mutual influence of total populations.

We shall discuss certain features of the behavior, for large values of t, of solutions of (2) subject to homogeneous Neumann boundary conditions,

$$\frac{\partial u}{\partial n} = 0 , \quad \frac{\partial v}{\partial n} = 0 \qquad \text{for } t > 0, \quad x \in \partial\Omega . \tag{3}$$

This boundary condition can be interpreted as an assumption that there is no migration across the boundary of Ω i.e. that both populations are confined to Ω.

Our motivation for studying (2) is two-fold. First, there is growing interest in the effects of mobility and spatial variation upon the classical ecological interactions (see e.g. [8] and [11]). Secondly, (2) is an example of a reaction-diffusion system the study of which is of increasing interest in various fields (cf [1], [6], [7], [12] and the article of Aronson in this collection).

The results we present in this paper are mathematical and their interest from an ecological point of view is only foundational. On the other hand they possess the virtue of generality and lack of ambiguity. For each of the three classical ecological interactions we discuss three main topics. 1. We consider the question of pointwise bounds for solutions of (2)-(3). We show that if the local growth rates become negative for large values of population densities (i.e. there is an ultimate limit to the population densities that can be supported by the environment) then for all non-negative and bounded initial values,

$$u(0,x) = u_0(x) \ , \ v(0,x) = v_0(x) \ , \ x \in \Omega \ , \tag{4}$$

the solution $(u(t,x) \ , \ v(t,x))$ is eventually contained in a distinguished subset of phase space (u-v space). This subset is a rectangle which is determined by the functions M and N and is independent of the particular initial values. 2. Our second result deals with the question of asymptotic stability of $(0,0)$, the equilibrium point corresponding to extinction of both populations. We show that if $(0,0)$ is asymptotically stable as a singular point of the vector field $(uM(u,v) \ , \ vN(u,v))$ then it is also asymptotically stable as a (constant) solution of (2)-(3). Thus spatial variability and diffusion does not alter the nature of the extinction state. 3. These first two results are valid for all non-negative values of the diffusion constants, d_1 and d_2. However the last question that we consider concerns when it makes sense to ignore spatial variation and confine attention to the ordinary differential equations. We show that when the rate of diffusion relative to the size of Ω is large compared to the strength of the interaction between the two populations then all solutions of (2)-(3) decay exponentially to spatially homogeneous solutions of ordinary differential equations having the same asymptotic limit sets as (1).

Basic to our discussion is the notion of an invariant rectangle for (2)-(3) to which we now turn.

2. INVARIANT RECTANGLES AND MAXIMAL FIELDS

We assume that the functions M and N are smooth and that Ω is a bounded open set with sufficiently regular boundary (cf. [3] and [5] for a somewhat more explicit statement of requirements). A rectangle Σ,

$$\Sigma = \{(u,v): a_1 < u < a_2 \ , \ b_1 < v < b_2\} \tag{5}$$

is said to be invariant for (2)-(3) if whenever

$$(u(0,x), \ v(0,x)) \in \bar{\Sigma} \ \text{ for all } \ x \in \Omega$$

it follows that $(u(t,x), v(t,x)) \in \Sigma$ for all $t > 0$, $x \in \Omega$. This notion is discussed in [2] and [3] where it is shown that for a rectangle to be invariant it is necessary and sufficient that the vector field $(u\,M(u,v), v\,N(u,v))$ not point out of Σ, i.e. $u = a_1 \Rightarrow u\,M(u,v) \geq 0$ for $b_1 < v < b_2$; $u = a_2 \Rightarrow u\,M(u,v) \leq 0$ for $b_1 < v < b_2$; $v = b_1 \Rightarrow v\,N(u,v) \geq 0$ for $a_1 < u < a_2$, etc. It is thus immediately clear that the positive quadrant,

$$\Sigma_0 = \{(u,v): 0 < u \ , \ 0 < v\}$$

is an invariant rectangle so that if u and v are initially non-negative then they remain so on the full interval of existence. This is a simple consequence of the form of equations (2); it is true for all smooth functions, M and N.

We define M^+ and N^+, the maximal functions relative to Σ_0 by

$$M^+(u,v) = \max[M(u,\theta): 0 \leq \theta \leq v]$$

$$N^+(u,v) = \max[N(\theta,v): 0 \leq \theta \leq u] \ .$$

In [3] we showed that M^+ and N^+ are locally Lipschitz continuous in Σ_0 so that the following differential equations have unique solutions:

$$\frac{du}{dt} = u\,M^+(u,v) \ , \quad \frac{dv}{dt} = v\,N^+(u,v) \ . \tag{7}$$

The following result is a special case of a theorem of [3].

Comparison Theorem. Let $u(t,x)$, $v(t,x)$ be a solution of (2)-(3) on $(0,\infty) \times \Omega$. Let $(U^+(t), V^+(t))$ be a solution of (7) on $(0,\infty)$. If $u(0,x) \leq U^+(0)$, $v(0,x) \leq V^+(0)$ for all x in Ω then $u(t,x) \leq U^+(t)$, $v(t,x) \leq V^+(t)$ for all $t \geq 0$, $x \in \Omega$.

3. ECOLOGICAL INTERACTIONS

For general background we refer to [10], [11] and [13]. For two species there are three classic interactions and these are determined by the signs of the partial derivatives,

$$M_v \equiv \frac{\partial M}{\partial v} \qquad N_u \equiv \frac{\partial N}{\partial u} .$$

In the predator-prey interaction the derivatives are of opposite sign:

$$M_v(u,v) < 0 , \quad N_u(u,v) > 0 \quad \text{for} \quad u > 0 , \ v > 0 . \qquad \text{P.}$$

Competition refers to the case when both derivatives are negative:

$$M_v(u,v) < 0 , \quad N_u(u,v) < 0 . \qquad \text{C.}$$

In symbiosis or mutualism both are positive

$$M_v(u,v) > 0 , \quad N_u(u,v) > 0 . \qquad \text{S.}$$

These are minimal assumptions which are imposed in virtually all discussions except that, for expository reasons, we have restricted our discussion to strict inequalities rather than allow the derivatives to vanish. Even in the context of ordinary differential equations many studies have shown the importance of imposing further conditions which reflect an ultimate limit to growth of one or both species. We will show that when dealing with the partial differential equation (2)-(3) such limits to growth are intimately related to pointwise bounds for solutions. It is interesting that for this most fundamental theoretical desiderata we are led to impose strict limits to growth on both species. Our specific assumptions are as follows.

Predator-Prey

PL 1. There is a $k_0 < 0$ such that $M(u,0) < 0$ for all $u > k_0$.

PL 2. There is a function ℓ such that $N(u,v) < 0$ for all $u > 0$ and $v > \ell(u)$.

Competition

CL 1. There is a $k_0 > 0$ such that $M(u,0) < 0$ for all $u > k_0$.

CL 2. There is a $\ell_0 > 0$ such that $N(0,v) < 0$ for all $v > \ell_0$.

Symbiosis

SL 1. There is a function k such that $M(u,v) < 0$ for $v > 0$ and $u > k(v)$.

SL 2. There is a function ℓ such that $N(u,v) < 0$ for $u > 0$ and $v > \ell(u)$.

SL 3. $k(v) = o(v)$ and $\ell(u) = o(u)$ for large values of their arguments.

These conditions need little explanation. In the case of PL 1 we are saying that even when there are no predators ($v = 0$), i.e. under the most favorable conditions for u, the environment does not allow growth once the density exceeds a critical value, k_0. Conditions CL 1 and CL 2 are to be interpreted similarly. In the case of PL 2 we notice that because of condition P an increase of u represents an enrichment of the environment for v. But PL 2 insures that no matter what the value of u the growth rate for v becomes negative once v becomes large enough. Conditions SL 1 and SL 2 are to be interpreted similarly. Now because of the smoothness of M and N we see that $N(u,\ell(u)) = 0$ for all $u > 0$, $\ell(u) > 0$. Hence,

$$N_u(u,\ell(u)) + N_v(u, (u))\ell'(u) = 0 .$$

Since $N_u > 0$ in conditions P and S and since $N_v(u,\ell(u)) \leq 0$ because of PL 2 and SL 2, it follows that $\ell'(u) > 0$. In a similar way we see that $k'(v) > 0$. This is merely a statement of the fact that in P and S an increase of u is advantageous for the growth of v. Similarly in Symbiosis an increase of v is advantageous for u. Condition SL 3 requires that this enhancement diminish for very large values of the densities.

We see then that in each interaction the following condition is satisfied.

(L) Limitation to Growth. There are nondecreasing, non-negative functions, k and ℓ, such that

$$u > k(v) \Longrightarrow M(u,v) < 0$$

and

$$v > \ell(u) \Longrightarrow N(u,v) < 0 .$$

We assume that k and ℓ are the smallest such functions. Moreover, both of the following two conditions are satisfied for large values of the argument:

$$k(v) = o(v) , \ \ell(u) = o(u) .$$

We now define the set B by

$$B = \{(u,v)\colon u \geq k(v) \text{ and } v \geq \ell(u)\} .$$

If condition L is satisfied then B is a nonempty, unbounded subset of the positive quadrant having nonempty interior. The set B need not be connected but it has one unbounded component which we refer to as B_∞. Because of L there is a unique point (K,L) such that if (u,v) is in B_∞ then $u \geq K$ and $v \geq L$. Figure 1 gives an example in the case of the predator-prey and the symbiosis interactions.

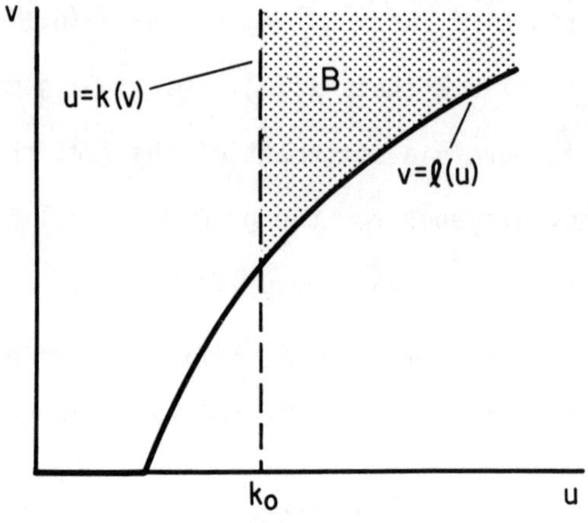

Figure 1 (a)

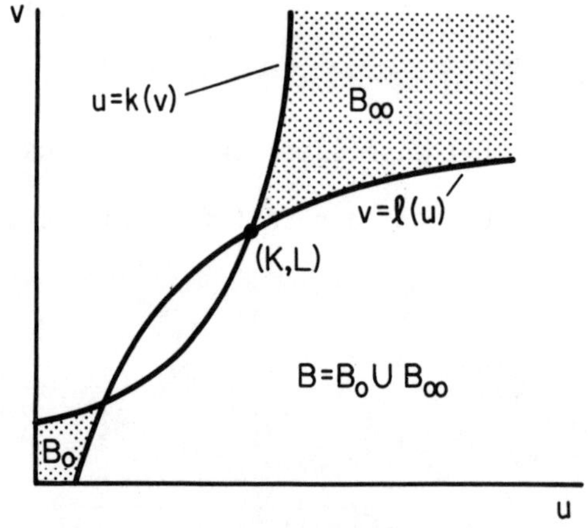

Figure 1 (b)

We now let $\Sigma(a,b) = \{(u,v): 0 < u < a \text{ and } 0 < v < b\}$.

Theorem 1. If M and N are smooth and condition L is satisfied then if $(a,b) \in B$ it follows that $\Sigma(a,b)$ is an invariant rectangle for (2)-(3).

Proof. Since $a \geq k(b)$ and since k is nondecreasing it follows that $a \geq k(v)$ for all $0 \leq v \leq b$. Hence, $aM(a,v) \leq 0$ for $0 \leq v \leq b$. Similarly, since $b \geq \ell(a) \geq \ell(u)$ for $0 \leq u \leq a$ it follows that $bN(u,b) \leq 0$ for $0 \leq u \leq a$. Hence, from our discussion in Section 2 it follows that $\Sigma(a,b)$ is invariant.

Thus if (u,v) is a solution of (2)-(3) then, since B is unbounded, we may find (a,b) in B such that $0 \leq u(0,x) \leq a$ and $0 \leq v(0,x) \leq b$ for all x in Ω. It follows from Theorem 1 that for all $t \geq 0$ and x in Ω we have

$$0 \leq u(t,x) \leq a \ , \quad 0 \leq v(t,x) \leq b \ .$$

But a stronger statement can be made and to prepare for this we first prove the following

Lemma 1. The maximal functions, M^+ and N^+, are both negative in the interior of B.

Lemma 2. B is an invariant set for the maximal field $(uM^+(u,v) \ , \ vN^-(u,v))$.

Proof. If (u,v) is in the interior of B then $u > k(v)$. Since k is nondecreasing it follows that $u > k(\theta)$ and hence, that $M(u,\theta) < 0$ for $0 \leq \theta \leq v$. Therefore, $M^+(u,v)$ is negative. That $N^+(u,v)$ is negative is seen in the same way. This proves Lemma 1. To see that B is invariant for the maximal field, i.e. invariant for the ODE (7), we point out that the boundary of B is made up of segments where either (i) $u \equiv 0$, (ii) $v \equiv 0$, (iii) $v = \ell(u)$, $\ell'(u) \geq 0$ or (iv) $u = k(v)$, $k'(v) \geq 0$. Now trajectories of (7) clearly cannot cross segments of type (i) or (ii).

In the interior of B, $v > \ell(u)$ and $N^+ < 0$. Moreover, since ℓ was chosen to be the smallest function satisfying the conditions of L it follows that $N^+(u, \ell(u)) = 0$. Hence, a trajectory of (7) cannot pass out of B across a segment of type (iii). The reader may find it helpful to consult Figure 2.

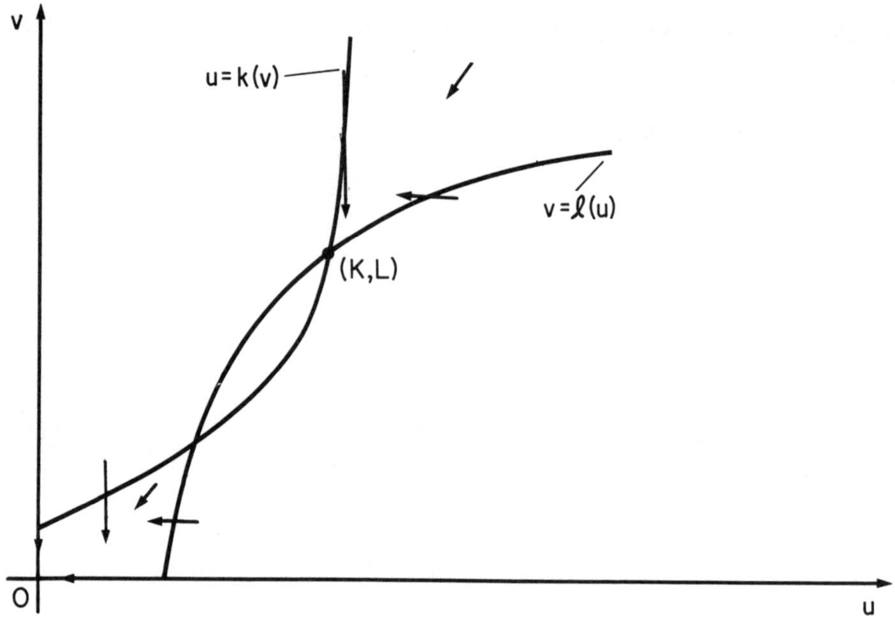

Figure 2. Maximal Field in B: Example of Symbiosis

A similar argument shows that trajectories of (7) cannot pass from the interior of B across a segment of boundary of type (iv). This completes the proof of Lemma 2. As an immediate corollary of Lemma 1 and 2 we have

Lemma 3. Every solution of the maximal ODE (7) which for $t = 0$ is in B_∞ converges as $t \to \infty$ to the point (K,L) which is a singular point of (7).

We now consider any smooth solution (u,v) of (2)-(3)-(4) on $[0,\infty) \times \Omega$ for which u_0 and v_0 are non-negative and bounded. As noted above we can find a point (a,b) in B_∞ such that

$$0 \leq u_0(x) \leq a \ , \quad 0 \leq v_0(x) \leq b \ , \quad x \in \Omega \ . \tag{8}$$

Now let $(U^+(t), V^+(t))$ be the solution of (7) such that

$$(U^+(0), V^+(0)) = (a,b). \tag{9}$$

From Lemma 3 it follows that

$$\lim_{t\to\infty} (U^+(t), V^+(t)) = (K,L) \tag{10}$$

From the comparison theorem of Section 2 we can then conclude

$$\limsup_{t\to\infty} u(t,x) \leq K \ , \quad \limsup_{t\to\infty} v(t,x) \leq L \ . \tag{11}$$

We thus have proved

Theorem 2. Let Ω be a bounded domain with sufficiently regular boundary. Let M and N be smooth and satisfy condition (L). Then every smooth solution of (2)-(3) which is initially non-negative remains so and satisfies (11), i.e. the values of the solution lie in $\Sigma(K+\epsilon \ , L+\epsilon)$ for all sufficiently large t.

We emphasize that K and L are determined only by the functions M and N.

Stability of Extinction. The origin $(0,0)$ is a singular point for (uM,vN) that corresponds to extinction of both populations. A simple calculation shows that the eigenvalues of the linearization of this field about $(0,0)$ are precisely

$$\lambda = M(0,0) \quad , \quad \mu = N(0,0)$$

The stability character of $(0,0)$. i.e. the signs of λ and μ, vary from model to model in published discussions of the ecological interactions. In some discussions $(0,0)$ is assumed a repeller, in others an attractor and in still others it is a saddle. But this refers to the stability character of $(0,0)$ as an equilibrium solution of the ODE (1). The constant $u \equiv 0$, $v \equiv 0$ is also a solution of the PDE (2)-(3) and it is natural to wonder about its stability as such a solution i.e. its stability with respect to spatially inhomogeneous perturbations. In general this is a rather subtle question but we can make one fairly general statement.

Theorem 3. If $\lambda = M(0,0)$ and $\mu = N(0,0)$ are both negative then there are numbers k_1, ℓ_1 such that if

$$0 \le u_0(x) \le k_1 \quad \text{and} \quad 0 \le v_0(x) \le \ell_1 \tag{12}$$

then

$$\lim_{t \to \infty} (u(t,x),\ v(t,x)) = (0,0) \tag{13}$$

uniformly in Ω. In fact

$$u(t,x) = O(\bar{e}^{\lambda t}) \quad \text{and} \quad v(t,x) = O(\bar{e}^{\mu t}) \tag{14}$$

Proof. Since $M(0,0)$ and $N(0,0)$ are negative and M and N are smooth we can find $k_1 > 0$, $\ell_1 > 0$ such that if (u,v) is in $\Sigma(k_1,\ell_1)$ then $M(u,v) < 0$ and $N(u,v) < 0$. Hence M^+ and N^+ are also negative in $\Sigma(k_1,\ell_1)$. If $(U^+(t),\ V^+(t))$ is the solution of (7) such that

$$(U^+(0),\ V^+(0)) = (k_1,\ell_1)$$

then $U^+(t) \downarrow 0$ and $V^+(t) \downarrow$ to 0 as $t \to \infty$. From the comparison theorem of Section 2 it follows then that (13) is a consequence of (12). Since

$M^+(0,0) = M(0,0) = \lambda$ and $N^+(0,0) = N(0,0) = \mu$ the relation (14) also follows from $u(t,x) \le U^+(t)$, $v(t,x) \le V^+(t)$. This completes the proof of Theorem 3.

4. CONVERGENCE TO SPATIAL HOMOGENITY.

In this section we specialize the results of our work in [5] to the system of ecology equations (1)-(2)-(3). We refer the reader to [5] for the proofs of Theorems 4 and 5.

It is natural to expect that if the spatial region Ω is small or if the rate of diffusion is large then spatial variations of solutions of (2)-(3) should be damped out. We can show that this intuitive expectation is indeed justified but to do so it turns out to be absolutely crucial that there is a bounded invariant rectangle for (2). If M and N satisfy Condition (L) then from Theorem 1 we know there are arbitrarily large invariant rectangles of the form

$$\Sigma = \{(u,v): 0 < u < a, \quad 0 < v < b\}.$$

Let f be the map $(u,v) \longrightarrow (u\,M(u,v),\ v\,N(u,v))$ and let df be its differential. Let S denote the supremum of the matrix norm of df as (u,v) ranges over Σ. The number S is thus a measure of the strength of the nonlinear interaction terms of (2) over Σ, a subset of phase space. Let λ_1 denote the smallest positive eigenvalue of $-\Delta$ in Ω with homogeneous Neumann conditions. Let $d = \min(d_1, d_2)$ and define

$$\sigma = \lambda_1 d - S.$$

For any solution of (2)-(3) we define their mean values by

$$\overline{u}(t) = \frac{1}{|\Omega|} \int_\Omega u(t,x)\, dx$$

$$\overline{v}(t) = \frac{1}{|\Omega|} \int_\Omega v(t,x)\, dx .$$

A sufficient condition for the damping of spatial variations is given by

Theorem 4. If $\sigma > 0$ then every solution of (2)-(3) which is initially confined to Σ satisfies the following:

$$\sup_{x\in\Omega} |u(t,x) - \overline{u}(t)| + \sup_{x\in\Omega} |v(t,x) - \overline{v}(t)| \leq C\, \overline{e}^{\sigma t} . \tag{15}$$

The constant C may be taken to be proportional to supremum of the gradient of the initial values of u and v.

Since λ_1 is inversely proportional to the nth root of the measure of Ω we see that σ will be positive when the ratio of diffusion coefficient to $|\Omega|$ is large relative to S, the strength of the nonlinear interaction. Now S depends upon the particular choice of invariant rectangle Σ. But from Theorem 3 we know that every bounded solution of (2)-(3) is eventually in $\Sigma(K + \epsilon, L + \epsilon)$ for arbitrary $\epsilon > 0$. Thus if S_0 is the supremum of $\|df\|$ over $\Sigma(K,L)$ and if $\sigma_0 = \lambda_1 d - S_0$ then as a consequence of Theorem 3 and 4 we have

Theorem 4A. If $\sigma_0 > 0$ and condition (L) is satisfied then every solution of (2)-(3) decays uniformly to its mean values according to (15) with $\sigma = \sigma_0$.

Now when σ is positive so that spatial dependence is not significant it becomes natural to ask whether the ODE (1) can replace the PDE (2). This is indeed the case but only if we are interested in asymptotic behavior. The mean values, $\overline{u}(t)$ and $\overline{v}(t)$, are not solutions of (1). In fact it is clear from (2) and (3) that

$$\frac{d\overline{u}}{dt} = \frac{1}{|\Omega|} \int_\Omega u\, M(u,v)\, dx \ . \tag{16}$$

But as we show in [5] it is a simple consequence of Theorem 4 that

$$\begin{cases} \dfrac{d\overline{u}}{dt} = \overline{u}\, M(\overline{u},\overline{v}) + p(t) \ , \ p(t) = \mathcal{O}(\overline{e}^{\,\sigma t}) \\ \dfrac{d\overline{v}}{dt} = \overline{v}\, N(\overline{u},\overline{v}) + q(t) \ , \ q(t) = \mathcal{O}(\overline{e}^{\,\sigma t}) \ . \end{cases} \tag{17}$$

Since $(\overline{u}(t), \overline{v}(t))$ is in Σ for all $t > 0$ it follows from a result of Markus [9] that the ω-limit set of $(\overline{u}(t), \overline{v}(t))$, $t > 0$, is contained in the ω-limit set of (1) in Σ. We summarize this as

Theorem 5. Under the conditions of Theorem 4 the limiting functions, $\overline{u}$ and $\overline{v}$, are solutions of the ODE (17), an exponentially decaying perturbation of the Kolmogorov equations (1). The asymptotic behavior of $(\overline{u},\overline{v})$ is completely determined by (1) in Σ.

Now if, as in Theorem 4A, we assume that σ_0 is positive then we can confine our attention to $\Sigma(K,L)$.

Theorem 6. If $\sigma_0 > 0$ and Condition (L) is satisfied then every solution of the PDE (2)-(3) decays uniformly and exponentially to spatially homogeneous functions $(u(t), v(t))$ whose ω-limit set is a subset of the ω-limit set of the Kolmogorov Equations (1) in the distinguished invariant rectangle $\Sigma(K,L)$.

Remark. On the basis of Theorem 6 it is clear that no essentially different behavior can be due to spatial variation and diffusion unless the rate of diffusion is weaker than the strength of the interaction.

REFERENCES

1. J. Auchmuty and G. Nicolis, Dissipative structures, catastrophes and pattern formation, Proc. Nat. Acad. Sci., U.S.A., 71 (1974), 2748-2751.

2. K. Chueh, C. Conley and J. Smoller, Positively invariant regions for systems of nonlinear diffusion equations, Ind. U. Math. J., to appear.

3. E. Conway and J. A. Smoller, A comparison theorem for systems of reaction-diffusion equations," submitted for publication.

4. E. Conway and J. A. Smoller, Diffusion and the predator-prey interaction, submitted for publication.

5. E. Conway, D. Hoff and J. Smoller, Large time behavior of solutions of systems of nonlinear reaction-diffusion equations, submitted for publication.

6. A. Gierer and H. Meinhardt, A theory of biological pattern formation, Kybernetik 12 (1972), 30-39.

7. N. Kopell and L. N. Howard, Plane wave solutions to reaction-diffusion equations, Studies in Appl. Math. LII (1973), 291-328.

8. S. Levin, Spatial patterning and the structure of ecological communities, pages 1-35 in Some Mathematical Questions in Biology, S. A. Levine (ed.): Lectures on Mathematics in the Life Sciences, Volume 8, 1976. American Mathematical Society, Providence.

9. L. Markus, Asymptotically autonomous differential systems, Contributions to the Theory of Nonlinear Oscillations, Vol. 3, 17-29. Annals of Mathematics Studies, No. 36, Princeton Univ. Press, Princeton, N. J., (1956).

10. R. M. May, Stability and Complexity in Model Ecosystems, Princeton Univ. Press, Princeton, N. J., (1973).

11. J. Maynard Smith, Models in Ecology, Cambridge, 1974.

12. J. Rauch and J. A. Smoller, Qualitative Theory of the Fitzhugh-Nagumo Equation, to appear in Advances in Math.

13. A. Rescigno and I. W. Richardson, The struggle for life: I. Two species, Bull. Math. Biophys., 29 (1967) 377-388.

Edward Conway
Department of Mathematics
Tulane University
New Orleans, Louisiana 70118

Joel Smoller
Department of Mathematics
University of Michigan
Ann Arbor, Michigan 48104

J W EVANS

Transition behavior at the slow and fast impulses

A general form of nerve axon partial differential equations given by FitzHugh [5] which includes the Hodgkin-Huxley equations is

$$\begin{aligned} \dot{V}_t &= V_{xx} + f(V,W) \\ W_t &= g(V,W) \end{aligned} \tag{1}$$

where V and W are functions of x (normalized distance) and t (normalized time) for all x and say for $t \geq 0$. For the Hodgkin-Huxley equations W is a 3-vector but following FitzHugh [6] we are going to let W be one dimensional and construct a model which has features in common with the Hodgkin-Huxley equations and which illustrates an application of some theorems in references 1, 2, 3, and 4. Following McKean [9] our model will be piecewise linear. We have included those features which best illustrate the nature of the results of the above papers. For this reason the model used here differs in some minor respects from that of McKean. The work of Rinzel and Keller [10] using McKean's piecewise linear model shows the benefits that accrue when tractable models of these complex systems are used.

We will be guided by the following. Consider the equations which result when $V_x = 0$ for all x and t (the space clamped equations). The result is that the x variable can be ignored and the ordinary differential equation

$$\begin{aligned} V_t &= f(V,W) \\ W_t &= g(V,W) \end{aligned} \tag{2}$$

is obtained. By injecting current into the space clamped axon the physiologist can vary V. And if desired the value of V can be taken to any fixed value and kept there (voltage clamp). There is a state where f and g vanish called the rest state which we take to be at $V = 0$ and $W = 0$. In experimental preparations if V is raised briskly from the rest state and kept at a fixed value in a certain range the physical correlate of f which is at first negative becomes positive and then approaches a fixed negative value. For us this means that as W moves according to the dynamics $W_t = g(V,W)$ for the fixed V the function $f(V,W)$ again for fixed V must pass from an initial negative value at $W = 0$ through a region of positivity in the V,W plane to a final negative value. This puts a requirement on our model that W must change when $W = 0$ and V is in a desired range and that it must approach a final equilibrium value for the fixed V. For the sake of simplicity we set

$$g(V,W) = V - W \tag{3a}$$

so that for fixed V we have that W tends to V and the line $W = V$ is the location of the $W_t = 0$ isocline. Now we set

$$f(V,W) = -V + H(W - a)H(V - 2W) \tag{3b}$$

where the Heaviside function H is defined by $H(x) = 1$ for $x \geq 0$ and $H(x) = 0$ for $x < 0$.

This definition of f introduces a triangular region between the lines $V = 1$, $W = a$ and $V = 2W$ in the $V > 0$ half-plane in which f is positive. For the present purposes we set $a = .02$. This is illustrated in Figure 1.

We now have

$$V_t = f(V,W) = -V + H(W - a)H(V - 2W)$$
$$W_t = g(V,W) = V - W \tag{4}$$

for the space clamped equations. If we examine the behavior of a solution to (4) which starts at $(V_0,0)$ where V_0 is positive and in the correct range we see that the solution passes through the triangular region of positive f on its way back to the rest state.

The traveling wave equations.

If we look for a solution to (1) of the form $V(t,x) = \varphi(x + \nu t)$, $W(t,x) = \psi(x + \nu t)$ we, of course, find that φ and ψ satisfy the ordinary differential equations

$$\nu\varphi' = \varphi'' + f(\varphi,\psi)$$
$$\nu\psi' = g(\varphi,\psi) \tag{5}$$

given by Hodgkin and Huxley [7] where ν is the velocity of propogation to the left of such a solution. These may be rewritten as

$$\varphi'' = \nu\varphi' - f(\varphi,\psi)$$
$$\psi' = \frac{1}{\nu} g(\varphi,\psi) \; . \tag{6}$$

In this form we see that there are very clear implications when $\nu > 0$ tends to zero or to infinity. The limiting behavior at $\nu = 0$ is that $\varphi'' = -f(\varphi,\psi)$ and ψ is such that $g(\varphi,\psi) = 0$ if it is defined at all. Now in our model and in the Hodgkin-Huxley equations this limiting behavior is defined and when $\varphi > 0$ and ψ is such that $g(\varphi,\psi) = 0$ it happens that $f(\varphi,\psi) < 0$ so that $\varphi'' > 0$ and φ increases without bound. As ν tends to infinity we see that the limiting behavior is that $\psi' = 0$ and that

$\varphi'' \gg 0$ when $\varphi' > 0$. In our model as in the Hodgkin-Huxley equations if V exceeds a certain fixed value $f(V,W)$ is negative for all W.

The result is that for very large ν any solution to the traveling wave equation (5) which starts out with $\varphi' > 0$ will rise to the region where $f < 0$ for all W before there is any appreciable change in W and then continue to increase without bound. The path traced out by φ,ψ in the V,W plane thus rises in our model above the line $V = 1$ before ψ exceeds the value a appearing in (4) and thus does not enter the region in the $V > 0$ half-plane where $f > 0$. The same qualitative behavior is observed in the Hodgkin-Huxley equations.

Of course we are interested here in bounded solutions of (5) and in particular in any solution which rises from and returns to the rest state. The remarkable thing to observe is that both at very low velocity and very high velocity ν the same behavior is observed of solutions to (5) which leave the rest state with increasing φ namely that they miss completely the region in the $V > 0$ half-space where $f > 0$ and thus increase without bound.

To study the behavior at intermediate ν we consider first the related linear equations

$$\begin{aligned} V_t &= V_{xx} - V \\ W_t &= V - W \end{aligned} \tag{7}$$

The corresponding traveling wave equation becomes

$$\begin{aligned} \varphi'' &= \nu\varphi' - \varphi \\ \psi' &= \frac{1}{\nu}(\varphi - \psi) \end{aligned} \tag{8}$$

or

$$\begin{pmatrix} \gamma' \\ \varphi' \\ \psi' \end{pmatrix} = \begin{pmatrix} \nu & -1 & 0 \\ 1 & 0 & 0 \\ 0 & \frac{1}{\nu} & -\frac{1}{\nu} \end{pmatrix} \begin{pmatrix} \gamma \\ \varphi \\ \psi \end{pmatrix} \tag{9}$$

where $\gamma \overset{\text{def}}{\equiv} \varphi'$.

The eigenvalues of the matrix in (9) are $\mu_1 \overset{\text{def}}{\equiv} \frac{\nu + \sqrt{\nu^2 + 4}}{2}$, $\mu_2 \overset{\text{def}}{\equiv} \frac{\nu - \sqrt{\nu^2 + 4}}{2}$ and $\mu_3 = -\frac{1}{\nu}$ and the corresponding eigenvectors are

$$Y_1 \overset{\text{def}}{\equiv} \begin{pmatrix} \mu_1 \\ 1 \\ 1/(1 + \nu\mu_1) \end{pmatrix}, \quad Y_2 \overset{\text{def}}{\equiv} \begin{pmatrix} \mu_2 \\ 1 \\ 1/(1 + \nu\mu_2) \end{pmatrix}, \quad Y_3 \overset{\text{def}}{\equiv} \begin{pmatrix} 0 \\ 0 \\ 1 \end{pmatrix}.$$

Now we can examine the solutions of (9). There is a one dimensional subspace of solutions tending to rest at $-\infty$ consisting of multiples of Y_1 and a two dimensional subspace of solutions tending to rest at ∞ consisting of the span of Y_2 and Y_3 . A close look at Y_1 shows that as ν tends to 0 the vector Y_1 tends to ${}^t(1,1,1)$ (where t denotes transpose) and as ν tends to ∞, Y_1 tends to ${}^t(\infty,1,0)$. The vector Y_2 moves from ${}^t(-1,1,1)$ as ν tends to 0 to ${}^t(0,1,1)$ as ν tends to ∞.

Let us try for a picture of this in 3 space. Assign the vertical axis to φ, the horizontal axis in the page to ψ and the axis perpendicular and out from the front of the page to γ. It is left to the reader to picture the positions of the stable and unstable subspaces as ν changes.

Now we want to add to this 3 dimensional picture since our $f(V,W)$ differs from $-V$ of the linear equations (7) by the amount 1 when $a \le W \le V/2$. We thus picture the wedge $a \le \psi \le \varphi/2$ in (γ,φ,ψ) space and the intersection of the subspace of multiples of Y_1 and the subspaces spanned by Y_2 and Y_3 with the boundary of this wedge. As noted before the portion of this wedge with $\varphi < 1$ is the region where $f > 0$.

If we take the trajectories of the solutions to the linear system (9) and translate by increasing the φ coordinate by 1 we, of course, have the trajectories solutions to (5) inside the wedge. See if you can visualize this.

We now give some numerical solutions. Recall that a solution rising from the rest point (0,0,0) travels in the direction Y_1 until it hits the wedge. Note that for low ν it misses the wedge passing to the right in our picture. In order to enter the wedge we must have $\psi \leq 2\varphi$ and from the values in Y_1 this gives $1/(1 + \nu\mu_1) \leq 2$ or $\nu \geq 1/\sqrt{2} \cong .7071$. Also, if at the point of entry (which occurs only when $\psi = a$) we have $\varphi \geq 1$ the φ coordinate will continue to increase without bound so if ν exceeds the value (where $1/(1 + \nu\mu_1) = a = .02$) of $9.8/\sqrt{2} \cong 6.9296$ there can be no bounded solution.

In Figure 2 the projection into the φ,ψ plane of the trajectories of the solutions to (5) rising from ${}^t(0,0,0)$ for $\nu = 816$, $\nu = .81713178$ and $\nu = .818$ are shown. The solution for $\nu = .816$ comes out of the wedge above the subspace spanned by Y_2 and Y_3 and thus has a positive multiple of $e^{\mu_1 y} Y_1$ as a component. The solution with $\nu = .81713178$ (within the limits of numerical error) leaves the wedge in the subspace spanned by Y_2 and Y_3 and thus returns to the rest state ${}^t(0,0,0)$. The solution with $\nu = .818$ comes out of the wedge below the subspace spanned by Y_2 and Y_3 and thus has a negative multiple of $e^{\mu_1 y} Y_1$ as a component so that it decreases without bound as y tends to infinity.

In Figure 3 trajectories corresponding to $\nu = 2.16$, $\nu = 2.1697425$ and $\nu = 2.17$ are shown. Here the opposite transitional behavior is observed: from below to inside (within the limits of error) to above the subspace spanned by Y_2 and Y_3. These are the only such transitions observed for the first exit from the wedge of trajectories leaving the rest point. Of course the first transition gives the "slow impulse" described for the Hodgkin-Huxley equations by Huxley [8] and computed by Rinzel and Keller [10] for the equations of McKean while the second transition gives the "fast impulse." Our attention is focused on the direction of these transitions as a result of a theorem in reference 4 which asserts that the transition type observed for the slow impulse (from above to below the surface of solutions returning to rest) implies that this impulse is unstable under a small perturbation of the initial conditions. The direction of the transition alone does not imply stability of the fast impulse but if the fast impulse is stable the direction has to be as it is.

In summary, we have constructed a simple model which reflects some of the behavior of the Hodgkin-Huxley equations under space clamped conditions and for the traveling wave equations. The model shows clearly the transitional behavior associated with the slow and fast impulses. This transitional behavior is linked to the stability of the impulses and in fact is sufficient to show that the slow impulse is unstable. In a paper in preparation with J. Feroe some numerical integration of the Hodgkin-Huxley equations dealing with these aspects is carried out as well as an exposition of the theoretical content of references 1, 2, 3, and 4.

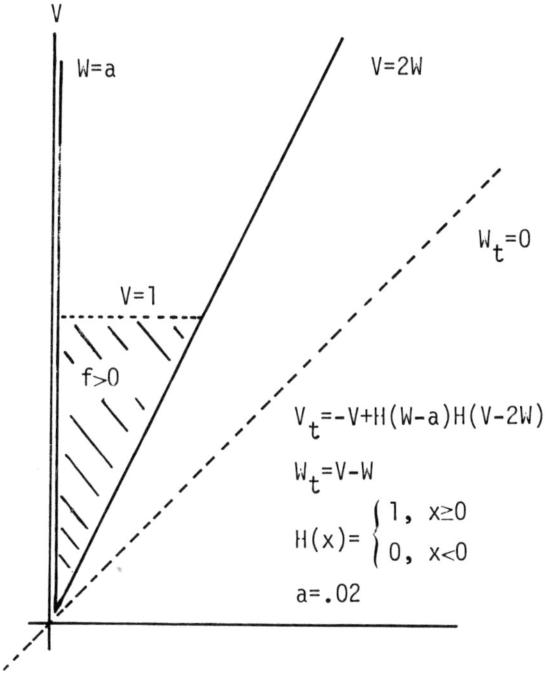

Figure 1

The location in the V,W plane of the triangular region where $f(V,W) > 0$ for $V > 0$ is shown along with the line $V = W$ where $W_t = 0$.

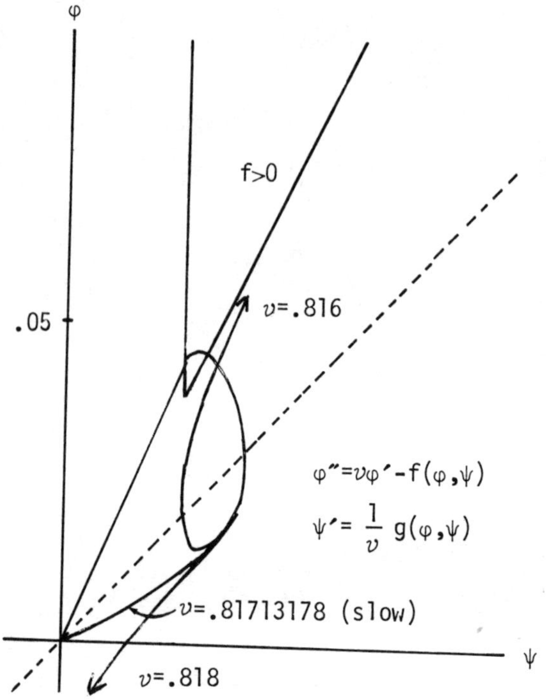

Figure 2

The projection into the φ,ψ plane of the trajectories rising from rest of solutions to the ordinary differential equation (5) are shown for $v = .816$, $v = .81713178$ and $v = .818$. A transition from above to inside to below the surface of solutions returning to rest is observed (see text).

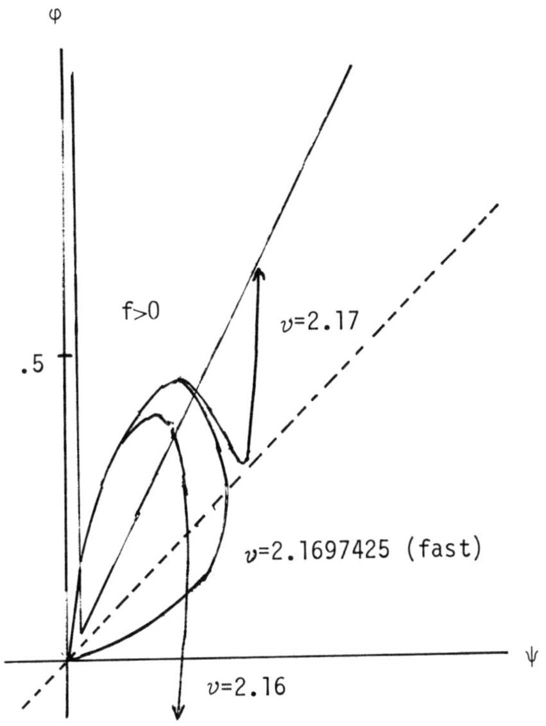

Figure 3

Trajectories rising from rest in the φ,ψ plane for $v = 2.16$, $v = 2.1697425$ and $v = 2.17$ are shown. Here the opposite transition behavior from that of Figure 2 is seen (see text).

REFERENCES

1. J. Evans, Nerve axon equations: I Linear approximations, Indiana Univ. Math. J. 21 (1972), 877-885.

2. J. Evans, Nerve axon equations: II Stability at rest, Indiana Univ. Math. J. 22 (1972), 75-90.

3. J. Evans, Nerve axon equations: III Stability of the nerve impulse, Indiana Univ. Math. J. 22 (1972), 577-593.

4. J. Evans, Nerve axon equations: IV The stable and the unstable impulse, Indiana Univ. Math. J. 24 (1975), 1171-1190.

5. R. FitzHugh, Mathematical models of excitation and propagation in nerve, Chapter 1 of Biological Engineering, H. P. Schwan, Ed., McGraw-Hill Book Company, Inc., New York, 1969.

6. R. FitzHugh, Impulses and physiological states in models of nerve membrane, Biophys. J. 1, (1961) 445-466.

7. A. L. Hodgkin and A. F. Huxley, A qualitative description of membrane current and its application to conduction and excitation in nerve, J. Physiol. 117 (1952) 500-544.

8. A. F. Huxley, Ion movement during nerve activity, Ann. N. Y. Acad. Sci. 81 (1959) 221-246.

9. N. P. McKean, Nagumo's equation, Quarterly J. Math. 4 (1970) 209-223.

10. J. Rinzel and J. Keller, Traveling wave solutions of a nerve conduction equation, Biophysical J., 19 (1973) 1313-1337.

John W. Evans
Department of Mathematics
University of California, San Diego
La Jolla, California 92093

P C FIFE *

Stationary patterns for reaction-diffusion equations

ABSTRACT

Patterns are defined to be stable stationary nonconstant solutions of the equations of reaction and diffusion. Several approaches are used to show the existence (or nonexistence) of patterns depending on one variable and defined on the entire real line. For a scalar equation, it is shown that there are essentially no patterns. For a system, small amplitude patterns, larger amplitude "peaks," and larger amplitude "plateaus" are treated. In all cases, stability is an important consideration. Applications to ecology and biophysics are mentioned.

1. INTRODUCTION

The equations of reaction and diffusion

$$\frac{\partial u}{\partial t} - D\Delta u = f(u), \qquad u = (u_1, u_2, \ldots, u_n) \tag{1}$$

(u and f are n-vectors, D a matrix, often taken to be diagonal and nonnegative) have been the object of a considerable number of studies in recent years, principally because of their actual and potential applicability to a variety of problems in population dynamics, biophysics, chemical physics, and chemical engineering.

Here we concentrate on the question of existence of stationary "patterns," which we define to be stable, stationary, nonconstant, bounded solutions of

* Sponsored in part by the United States Army under Contract DAAG29-75-C-0024 and in part by the National Science Foundation under Grant MPS-74-06835-A01.

(1). (We do not consider "moving patterns" here, such as those studied by Howard and Kopell, Winfree, and others.) Interest in such stationary solutions, sometimes called dissipative structures, was aroused by the work of Turing [27] in the 1950's, and by that of Gmitro and Scriven [15], Prigogine and Nicolis [22] and others in the 1960's and later. Particular interest in them has been occasioned by their possible role in reflecting the corresponding phenomena of pattern formation in developing organisms and in ecological communities. However, it should be emphasized that for most contexts in which (1) has been used to model a biological phenomenon (signal dynamics on nerve axons excepted), the model is highly idealized and accounts imperfectly for only part of the processes actually occurring to produce the phenomenon.

Except for remarks at the end of the paper, we restrict attention to solutions depending on only one space variable. This is necessary for some of our results, merely convenient for others. Possible extensions to many-dimensional problems are noted in section 4.

We are particularly interested in patterns which are not elicited by boundary effects, and for this reason further focus attention on patterns defined on the entire real line. Again in the last section, some discussion will be given about corresponding results on bounded domains.

The above definition of pattern requires that it be stable, with reference to the evolution system (1). For some results given here, this means C^0-stability, which we define as follows: a bounded stationary solution $\varphi(x)$ of (1) is C^0-stable if, given any $\varepsilon > 0$, there is a δ such that every solution $u(x,t)$ of (1) defined for $x \in \underset{\sim}{R}$, $t \geq 0$, satisfying $|u(\cdot,0) - \varphi|_0 \equiv \sup_{x \in \underset{\sim}{R}} |u(x,0) - \varphi(x)| < \delta$ also satisfies $|u(\cdot,t) - \varphi|_0 < \varepsilon$ for all $t \tilde{>} 0$.

In other cases, we are unable to prove C^0-stability, and instead give stability arguments based on an analysis of the spectrum of the linearization of the right side of (1) about the stationary solution in question. In still other cases, we give less rigorous heuristic arguments favoring stability.

Variations of the above reaction-diffusion problems, such as those obtained by allowing the diffusion terms to be nonlinear, and/or first derivative terms to enter the equation, are important in some applications, but will not be discussed here. Othmer [21] has discussed (among other things) standing oscillation solutions of (1), which bear the same relation to x-independent oscillatory solutions as do ours to constant solutions.

The paper is concerned with the mathematical analysis of patterns; experimental and numerical simulation results are not mentioned except in passing. Nor do we go into much detail about applications in the various fields mentioned above. For an extended discussion of the relevance of patterns in ecology, see (for example) the article by Levin [18].

The plan of the paper is as follows.

1. Introduction
2. The scalar case, $n = 1$
3. Systems, $n > 1$
   - A. Small amplitude patterns
   - B. Larger amplitude patterns
     1. Peaks
     2. Plateaus
4. Discussion

Throughout, we assume that f is a continuously differentiable function of its argument.

I gratefully acknowledge helpful discussions with D. H. Sattinger and with K. Kirchgässner, who convinced me that there exist quasiperiodic solutions of (14), and that only the cases shown in Figure 4 occur.

2. THE SCALAR CASE, n = 1

In this case, D is a scalar, which we take to be unity. The equation of one-dimensional patterns, then, is simply

$$u_{xx} + f(u) = 0, \quad x \in \underset{\sim}{R} . \tag{2}$$

This equation is well-known and easily solved by a phase-plane analysis or by the argument below. Furthermore, the stability of its bounded solutions can be tested by means of a technique due to Aronson and Weinberger [1].

The equation has a first integral, obtained by multiplying by the integrating factor u_x :

$$\frac{1}{2}(u_x)^2 + F(u) = E = \text{const} , \tag{3}$$

where

$$F(u) \equiv \int_0^u f(s)ds .$$

If one draws the graph of the "potential" $V = F(u)$, as illustrated in Fig. 1, then there is a one-one correspondence between the nonconstant solutions of (2) (modulo translation of the independent variable and reversal of its sign, which are always possible) and horizontal line segments in the (u,V) plane whose finite endpoints lie on the given curve, and which otherwise lie strictly above the curve. The projection of such a segment onto the u-axis is the range of the corresponding solution, and the ordinate of the segment is the constant E (its "energy") in (3). Given such a segment,

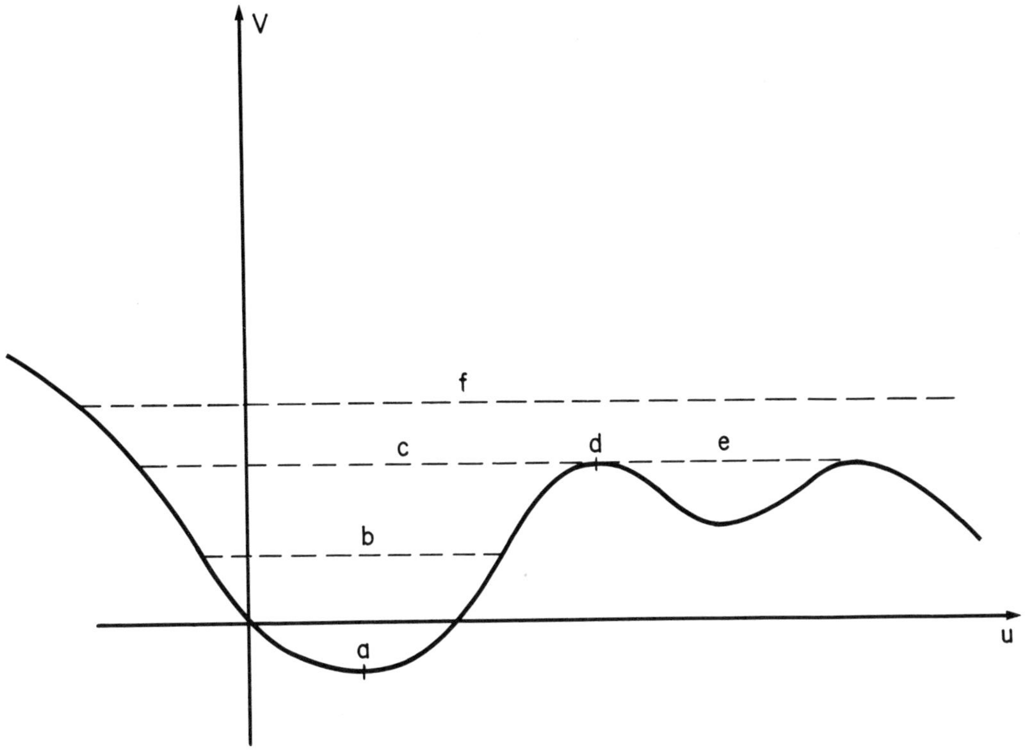

Figure 1.

The potential function $V = F(u)$. Horizontal lines (and points) such as a - f correspond to solutions of (2).

the solution $u(x)$ can be obtained by further integrating (3). On each interval of monotonicity, we obtain

$$x - x_0 = \pm 2^{-\frac{1}{2}} \int_{u_0}^{u} (E - F(s))^{-\frac{1}{2}} ds \ , \tag{4}$$

where u_0 is an arbitrary number in the interior of the (known) range, and x_0 is arbitrary. This excludes the constant solutions of (2); but they, of course can be identified with the points of zero slope in Figure 1, since then $F'(u) = f(u) = 0$.

Figure 1 illustrates six solutions: (a) and (d) are constant ones; (b) is a periodic solution; (c) is a solution attaining a minimum at a single finite value of x, and approaching its supremum as $x \to \pm\infty$; (e) is a monotone solution approaching different limits as $x \to \pm\infty$; and (f) is a solution bounded from below, but unbounded above. These properties can be deduced from (4). For example, in cases (c) and (e), the fact that $|x| \to \infty$ as u approaches its supremum u_m follows from the fact that $E - F(s) \le C(u_m - s)^2$ for s near u_m, so that the integral in (4) is unbounded as $u \uparrow u_m$. In the case (b), a similar analysis shows that the maximum and minimum are obtained for finite values of x. Then the uniqueness of initial value problems for (2), together with the sign reversibility of x, show the solution must be periodic.

With all possible solutions of (2) known, we now consider their C^0-stability as solutions of (1), which in our present case is

$$u_t - u_{xx} - f(u) = 0 . \qquad (5)$$

The result is that very few are stable.

Lemma 1: Every bounded nonconstant solution of (2) which attains a maximum or minimum at a finite value of x is C^0-unstable.

Proof: Let φ be such a solution. Assume it has a minimum $\varphi = m$ at $x = 0$ (the argument for the case of a maximum is the same). Then it corresponds to a segment I, such as that in Figure 2, whose left endpoint is on the curve $V = F(u)$ at a point $u = m$ such that $F'(m) < 0$. Let E be the ordinate of the segment I (the energy of φ), and for positive δ, let I_δ be the segment with energy $E + \delta$ overlapping I. Let φ_δ be the corresponding (possibly unbounded) solution of (2). For small enough δ, it will attain a minimum $m_\delta < m$, where $F(m_\delta) = E + \delta$. Furthermore

$$\lim_{\delta \downarrow 0} m_\delta = m . \qquad (6)$$

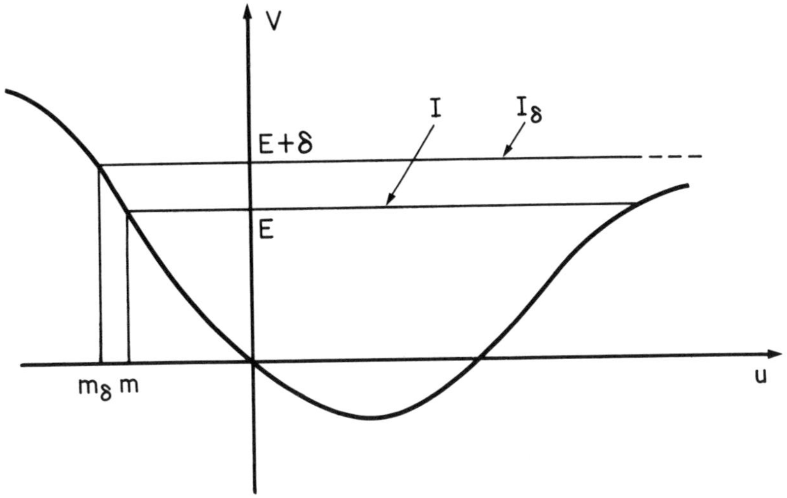

Figure 2.

By translation of the independent variable, if necessary, we may assume that φ_δ also attains its minimum at $x = 0$.

Let X_δ be the largest interval on the x-axis containing the origin, on which $\varphi_\delta(x) \le \varphi(x)$. Since I_δ is longer than I at both ends, φ_δ will assume values greater than $\sup_x \varphi(x)$. Therefore X_δ is finite. Let

$$\overline{\varphi}(x) = \begin{cases} \varphi_\delta(x), & x \in X_\delta ; \\ \varphi(x), & x \notin X_\delta . \end{cases}$$

We shall need the following fact, whose proof, being technical, is deferred until later.

<u>Lemma 2</u>:

$$\lim_{\delta \downarrow 0} |\varphi - \overline{\varphi}|_0 = 0 \quad .$$

Locally, $\overline{\varphi}$ is the minimum of two solutions, so will itself be a supersolution of (5). Let $\overline{u}(x,t)$ be the solution of (5) satisfying $\overline{u}(x,0) = \overline{\varphi}(x)$. By a result of Aronson and Weinberger [1], $\overline{u}(x,t)$ is decreasing in t, pointwise in x. (The proof in [1] was for the case $\varphi \equiv$ const, but works as well in our case.)

Let $[0,T)(T \leq \infty)$ be the maximum interval of existence of $\overline{u}$. By the decreasing property, $\lim_{t \uparrow T} \overline{u}(x,t) \equiv \psi(x)$ exists and, where $\psi > -\infty$, is a solution of (2) (see [1]).

If ψ is a solution for all x, then it must satisfy $\psi(x) \leq \sup \varphi$ and must attain values $\leq m_\delta < m$. This follows from the maximum principle and the fact that $\overline{\varphi}(x) \leq \varphi(x)$. But it may be seen that any horizontal segment or point in the $u - V$ plane corresponding to such a solution must lie entirely to the left of the value $u = m - \eta$, for some positive η independent of δ. Thus for each x,

$$\overline{u}(x,t) \leq m - \eta \tag{7}$$

for large enough t where $\overline{u}$ is defined.

On the other hand, if ψ is not a solution for all x, then $\psi = -\infty$ for some x, so in any case, (7) holds for at least some values of x and large enough t. (It can be shown, in fact, that either $\psi \equiv -\infty$, or $\psi \equiv$ const, and that $\overline{u} \to \psi$ as $t \to \infty$ uniformly on bounded sets. But we shall not use these results.)

Now let u be any solution of (5) satisfying $u(x,0) \leq \overline{\varphi}(x)$. By the maximum principle, $u(x,t) \leq \overline{u}(x,t)$, so there are some values of x for which $u(x,t) \leq m - \eta$ for all large enough t where u is defined. For such t, we have $|u(\cdot,t) - \varphi|_0 \geq \eta > 0$. But in view of Lemma 2, we may choose $|u(\cdot,0) - \varphi|_0$ arbitrarily small. Therefore φ is unstable, and the theorem is proved.

Besides the unstable ones covered in the theorem, the only nonconstant bounded solutions of (2) are the monotone ones, corresponding to (e) in Figure 1. These can be thought of as zero velocity travelling wave solutions in the sense of [13], where their global stability was proved. On the other hand, their existence depends on a special property of the curve F in Figure 1, namely that two adjacent local maxima have the same height. A slight change in f will destroy this property, so the monotone solutions are structurally unstable.

This exhausts the possible bounded solutions of (2), except for the constant ones. Though they are not candidates for being patterns, we include a discussion of their stability for the sake of completeness. As it turns out, a constant solution $u \equiv u_0$ of (2) is C^0-stable if and only if it is stable as a solution of the equation

$$\frac{du}{dt} = f(u) \quad . \tag{8}$$

Suppose first that u_0 is unstable as a solution of (8). Then there exists a neighborhood N of u_0, and solutions $u(t)$ of (8) with $u(0)$ arbitrarily close to u_0, but with $u(t) \notin N$ for some t. Such a function $u(t)$ is also a solution of (5), however. Therefore u_0 is C^0-unstable as a solution of (5) as well. Next, suppose u_0 is stable as a solution of (8). Then for each $\varepsilon > 0$, there is a $\delta(\varepsilon) > 0$ such that solutions of (8) with $|u(0) - u_0| \leq \delta$ exist for all time and satisfy $|u(t) - u_0| < \varepsilon$. Let $u(x,t)$ be a solution of (5) with $|u(\cdot,0) - u_0| \leq \delta$. By the maximum principle, it is bounded pointwise above and below, respectively, by $u_+(t)$ and $u_-(t)$, which are solutions of (8) assuming initial values $u_{\pm}(0) = u_0 \pm \delta$. Therefore $|u(\cdot,t) - u_0| < \varepsilon$, and $u(x,t)$ is stable.

Altogether, we have the following theorem.

Theorem: The only stable solutions of (2) are the strictly monotone ones, and the constant ones which are stable as solutions of (8). Moreover, the strictly monotone ones are structurally unstable.

Proof of Lemma 2: First, consider the case when φ is periodic. Then geometrical considerations show that the length of X_δ will not surpass twice the wave-length of φ, so remains bounded independently of δ. Then the continuity of solutions of initial value problems with respect to initial data easily implies $\lim_{\delta \downarrow 0} |\bar{\varphi}(x) - \varphi(x)| = \lim_{\delta \downarrow 0} (\varphi(x) - \varphi_\delta(x)) = 0$, uniformly in X_δ. But outside X_δ, $\bar{\varphi} \equiv \varphi$, so the conclusion follows.

The only other case is when φ approaches a limit asymptotically as $|x| \to \infty$, and $\varphi' > 0$ for $x > 0$. We treat this case as follows. By assumption, $F'(m) = f(m) < 0$. Let $\kappa = \frac{1}{2} f(m) < 0$, and restrict δ to be so small that $f(u) < \kappa$ for $m_\delta \le u \le m$. Then

$$
\begin{aligned}
\varphi_\delta(x) &= \varphi_\delta(0) + \int_0^x \int_0^s \varphi_\delta''(\xi)d\xi ds \\
&= \varphi_\delta(0) - \int_0^x \int_0^s f(\varphi_\delta(\xi))d\xi ds \\
&\ge m_\delta + \frac{1}{2} |\kappa| x^2 ,
\end{aligned}
$$

for x such that $m_\delta \le \varphi_\delta(x) \le m$. Let $\omega_\delta^2 = 2(m - m_\delta)/|\kappa|$, so that $m_\delta + \frac{1}{2}|\kappa|\omega_\delta^2 = m$. Then $\varphi_\delta(\omega_\delta) \ge m$. If we define π_δ as the first value of $x > 0$ at which $\varphi_\delta(x) = m$, then $\pi_\delta \le \omega_\delta$. It follows that

$$\lim_{\delta \downarrow 0} \pi_\delta = 0 . \tag{9}$$

Let $\psi_\delta(x) \equiv \varphi_\delta(x + \pi_\delta)$. Then $\psi_\delta(0) = m = \varphi(0)$, and from (3), for $x > 0$, $\psi_\delta' = [2(E + \delta - F(\psi_\delta))]^{\frac{1}{2}}$, whereas $\varphi' = [2(E - F(\varphi))]^{\frac{1}{2}} < [2(E + \delta - F(\varphi))]^{\frac{1}{2}}$. It follows that $\psi_\delta(x) \geq \varphi(x)$, for $x > 0$.

We now have, for $x \geq \pi_\delta$,

$$\begin{aligned}\varphi(x) - \varphi_\delta(x) &= \varphi(x) - \varphi(x - \pi_\delta) + \varphi(x - \pi_\delta) - \psi_\delta(x - \pi_\delta) \\ &\leq \varphi(x) - \varphi(x - \pi_\delta) .\end{aligned}$$

By the uniform continuity of φ and (9), we have that $\varphi(x) - \varphi_\delta(x) \leq \sigma_\delta \downarrow 0$ as $\delta \to 0$, for $x \geq \pi_\delta$. By symmetry, the same is true for $x \leq -\pi_\delta$ as well, and by the continuity of φ and φ_δ, we may extend the relation to the interior interval $[-\pi_\delta, \pi_\delta]$.

On X_δ we therefore have

$$|\overline{\varphi}(x) - \varphi(x)| = \varphi(x) - \varphi_\delta(x) \leq \sigma_\delta , \tag{10}$$

whereas outside X_δ, we have $|\overline{\varphi}(x) - \varphi(x)| = 0$. This completes the proof.

3. SYSTEMS (n > 1)

A. Small amplitude perturbations.

Let us consider systems (1) in which $f(0) = 0$, so that $u \equiv 0$ is a solution. It has been observed in the past that this trivial solution may be unstable with respect to certain perturbations which are not constant in x, while remaining stable with respect to constant (uniform) perturbations. Another way of saying this is that the trivial solution $u \equiv 0$ may be unstable, whereas the point $u = 0$ is a stable critical point of the associated kinetic equations

$$\frac{du}{dt} = f(u) . \tag{11}$$

In such cases, the effect of introducing diffusion and x-dependence is therefore to destabilize the system, somewhat contrary to one's intuition of diffusion as a stabilizing influence. An argument was advanced by Segel and Jackson [25] to partially explain this paradox. If one can characterize some of the components of u as "stabilizing" in some sense, and others as "destabilizing," and if the stabilizers diffuse more rapidly than the destabilizers, then a very small initial nonuniform concentration of both (this being the perturbation of the trivial solution) may result in the stabilizers diffusing away, their effect thereby diminishing. This would happen to some extent not only with the initial concentration of stabilizers, but also with stabilizers produced later through the reaction process. In this way, without as much counteracting influence, the destabilizers could take over, and the solution could grow to finite size. For this to happen, it is of course necessary that the diffusion coefficients not all be the same. In the example in [25], $n = 2$ and the system represented predator-prey equations, the predators being stabilizers and the prey destabilizers. Segel and Levin [26] have pursued similar problems and have shown that after instability sets in, one can expect solutions of (1) with small initial data sometimes to evolve into patterned solutions.

We consider the system (1), and assume the function f depends on a real parameter: $f = f(u,\lambda)$. We also assume that there is a critical number λ_{cr} such that for $\lambda < \lambda_{cr}$, the trivial solution of (1) is stable, in a sense to be explained below, whereas for $\lambda > \lambda_{cr}$, it is not. The new instability is with respect to periodic perturbations with wave length in a certain range, bounded above and below (the meaning of this is given below). In particular, the zero solution remains stable to constant perturbations, so the instability is of the type indicated in the previous paragraph. Under certain additional conditions, the effect of this is that periodic patterns of small amplitude arise for λ in a right-hand neighborhood of λ_{cr}. In fact, we shall show how bifurcation theory may be used to construct a two-parameter family (the parameters being amplitude and wave-length) in a neighborhood of λ_{cr}. The question of their stability will be taken up later.

Our assumptions need to be made precise. The source term f will be assumed twice continuously differentiable for u near the origin; we write A_λ for the Jacobian $\frac{\partial f}{\partial u}(0,\lambda)$. For simplicity, we assume the dependence of A_λ on λ is linear: $A_\lambda = A + \lambda B$, for some matrices A and B. The linear problem associated with (1),

$$V_t = DV_{xx} + AV + \lambda BV ,$$

has solutions of the form $V = \Phi \exp[ikx + \sigma t]$ for arbitrary real wave number k. Here Φ will be an eigenvector of the matrix

$$H(\lambda,p) \equiv -pD + A + \lambda B \quad (p = k^2) ,$$

with corresponding eigenvalue σ.

<u>Assumption 1</u>: There is a number λ_{cr} and a number $\delta > 0$, such that for $|\lambda - \lambda_{cr}| < \delta$ and for all $p \geq 0$, $H(\lambda,p)$ has a unique eigenvalue with largest real part, and it is real and simple. Denote it by $\sigma_1(\lambda,p)$, and let $I_\lambda = \{p : \sigma_1(\lambda,p) \geq 0\}$. Then

$$I_\lambda = \begin{cases} \emptyset \text{ for } \lambda_{cr} - \delta < \lambda < \lambda_{cr} , \\ \{p_0\} \text{ for } \lambda = \lambda_{cr} , \text{ where } p_0 > 0 , \\ \text{a bounded interval of positive length for } \lambda_{cr} < \lambda < \lambda_{cr} + \delta . \end{cases}$$

This assumption about $\sigma_1(\lambda,p)$ is depicted in Figure 3.

For simplicity we take $\lambda_{cr} = 0$ from now on (except in the example below).

It follows from Assumption 1 that the matrix $H_0 = H(0,p_0)$ has zero as a simple eigenvalue. It's nullspace is therefore spanned by some real null-vector Φ, and the nullspace of the adjoint matrix is spanned by some real vector Ψ.

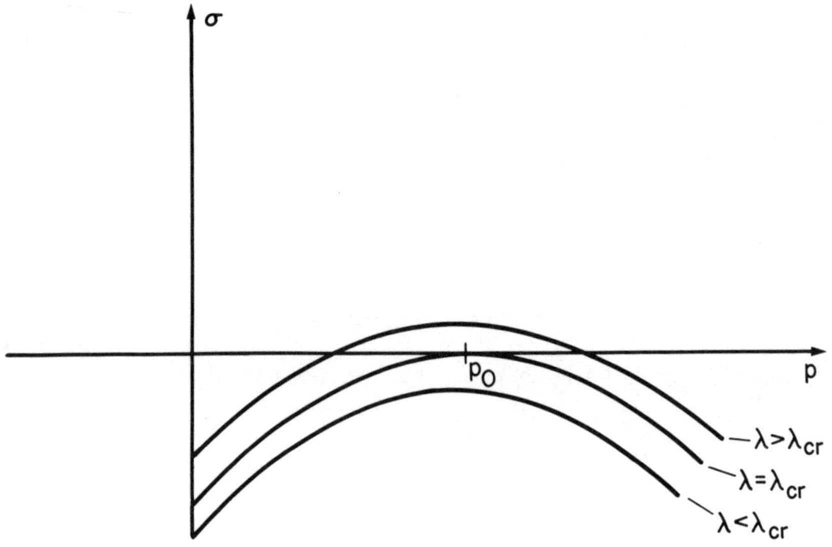

Figure 3.
Typical dependence of the largest eigenvalue σ_1 on λ and p, when Assumption 1 is valid. The curves are for λ constant.

<u>Assumption 2</u>: $\Psi B \Phi \neq 0$.

<u>Example</u>: Let $n = 2$,

$$A = \begin{pmatrix} 0 & -2 \\ 2 & -1 \end{pmatrix}, \quad B = \begin{pmatrix} 1 & 0 \\ 0 & 0 \end{pmatrix}, \quad \text{and} \quad D = \begin{pmatrix} 1 & 0 \\ 0 & 2 \end{pmatrix}$$

Then Assumption 1 is satisfied with $p_0 = \sqrt{2} - \frac{1}{2}$, $\lambda_{cr} = \frac{1}{2}(4\sqrt{2} - 1)$. In fact, the eigenvalues σ of $H(\lambda,p)$ are the roots of the equation $\sigma^2 + \sigma(3p + 1 - \lambda) + Q(p,\lambda) = 0$, where $Q(p,\lambda) = (p - \lambda)(2p + 1) + 4$. It can be shown that for small $|\lambda - \lambda_{cr}|$, the two roots are real and distinct for all $p \geq 0$, and the lesser of the two is negative and bounded away from zero. The greater one, $\sigma_1(p,\lambda)$, can be shown by the quadratic formula to satisfy

$$\sigma_1 \gtreqless 0 \quad \text{according as} \quad Q(p,\lambda) \lesseqgtr 0 .$$

But $Q(p,\lambda)$ is quadratic, and has a minimum with respect to p for each λ. Its minimum is 0 (at $p = p_0$) for $\lambda = \lambda_{cr}$, is positive when $\lambda < \lambda_{cr}$, and is negative when $\lambda > \lambda_{cr}$. Thus the set of values of p for which $\sigma_1 > 0$ is empty for $\lambda < \lambda_{cr}$, and is a finite interval for $\lambda > \lambda_{cr}$.

The basic result is the following.

Theorem: Assume

$$f(u,\lambda) = (A + \lambda B)u + G(u) , \tag{12}$$

where G is continuously differentiable near $u = 0$, and

$$|G(u)| \leq C|u|^2 \tag{13}$$

for $|u| < 1$. Let A, B, and D satisfy Assumptions 1 and 2. Let $k_0 = \sqrt{p_0}$ be the "initial" wave number. Then there exists a two-parameter family of periodic functions $u(x;\varepsilon,\kappa)$, and a scalar $\lambda(\varepsilon,\kappa)$, both continuous in ε and κ and defined for small $|\varepsilon|$, $|\kappa|$, such that

(i) u and λ satisfy

$$Du_{xx} + (A + \lambda B)u + G(u) = 0, \quad x \in \underset{\sim}{R}, \tag{14}$$

for each fixed ε and κ,

(ii) u is periodic in x with wave-length (period) $2\pi/k$ where $k = k_0 + \kappa$,

(iii) $\int_0^{2\pi/k} u(x;\varepsilon,\kappa)\cdot\Phi \cos kx dx = \varepsilon$,

(iv) $\lambda(0,0) = 0, \quad u(x;0,0) \equiv 0$.

We shall sketch the proof. The first step will be to give the problem a more convenient formulation. By using the new variable $\bar{x} = x/k$, we normalize the wave-length of our solutions to be 2π, but at the same time introduce an extra parameter $k^2 = p$ into the equations (14), which become

$$pDu_{\bar{x}\bar{x}} + (A + \lambda B)u + G(u) = 0 . \tag{15}$$

Henceforth we shall omit the bars on the x's. For $p = p_0$, $\lambda = 0$, the corresponding linear equation

$$p_0 DU_{xx} + AU = 0$$

has exactly two linearly independent solutions bounded for all x : $\varphi = \Phi \cos x$ and $\varphi' = -\Phi \sin x$. This follows from Assumption 1. We normalize our solutions so that $(\varphi,\varphi) \equiv \int_0^{2\pi} |\varphi|^2 dx = |\Phi|^2 \pi = 1$. Similarly, the linear adjoint problem has two linearly independent solutions $\psi = \Psi \cos x$ and $\psi' = -\Psi \sin x$.

Let P_1, P_2 be L_2-orthogonal projections onto span $\{\psi\}$ and span $\{\psi'\}$ respectively; for example, $P_1 f = (f,\psi)\psi$. Also let $Q = \text{id} - P_1 - P_2$. Let Z be the Banach space of continuous 2π-periodic functions with values in $\underset{\sim}{R}^n$, C^k the space of $\underset{\sim}{R}^n$-valued functions with bounded continuous derivatives of order k, $X = Z \cap C^2$ and Y the L_2 orthocomplement of φ and

φ' in X. We norm X and Y by the usual C^2 norm, and Z by the C^0 norm.

Then it is standard that the differential operator

$$L = p_0 D \frac{d^2}{dx^2} + A ,$$

restricted to Y, is one-one from Y onto QZ, so that L^{-1} exists as a bounded operator from QZ to Y.

Since $X = \text{span}\{\varphi, \varphi'\} \oplus Y$, any solution in X can be written as $u = \alpha\varphi + \beta\varphi' + v = \Phi(\alpha \cos x - \beta \sin x) + v$, $v \in Y$. Furthermore since (15) is autonomous, any solution remains a solution when subjected to an arbitrary translation of the independent variable. In fact, such a translation does not remove v from Y. We suppose our solutions have all been so translated in such a way that $\beta = 0$; we then let $\varepsilon = \alpha$ and call it the "amplitude" of the solution. In short, we seek solutions in the form

$$u = \varepsilon(\varphi + w), \quad w \in Y. \tag{16}$$

With the two final definitions $F(u,\varepsilon) = \varepsilon^{-1}G(\varepsilon u)$ (recall (13)) and $q = p - p_0$, we are ready for the reformulation. Setting (16) into (15) and using the above definitions, we see that the problem is to find a function $w \in Y$ satisfying

$$Lw + \lambda B(\varphi + w) + qD(\varphi + w)'' + \varepsilon F(\varphi + w,\varepsilon) = 0 . \tag{17}$$

We rewrite this as $Lw + R(w;\lambda,\varepsilon,q) = 0$, where for each (λ,ε,q), R is an operator from Y into Z. Applying projections Q, P_1, and P_2, we see that (17) is equivalent to the three equations

$$Lw + QR(w;\lambda,\varepsilon,q) = 0 , \tag{18a}$$

$$P_1R(w;\lambda,\varepsilon,q) = 0 \ , \tag{18b}$$

$$P_2R(w;\lambda,\varepsilon,q) = 0 \ . \tag{18c}$$

With the aid of Assumption 2 and the implicit function theorem, the first two equations (18a,b) may be solved uniquely for $w(\varepsilon,q)$ and $\lambda(\varepsilon,q)$ for small (ε,q). It remains to be shown that this solution will automatically satisfy (18c). Here we use the following simple invariance property of our problem. The operators L, R, Q, and $L^{-1}Q$ preserve the class of even functions. This fact is immediate for L, R, and Q, and not too difficult for $L^{-1}Q$. It follows that (18a-b) may also be solved in the class of even functions, so by uniqueness, the original solution must be even. But P_2, being orthogonal projection onto a space of odd functions, annihilates even functions. Therefore (18c) is satisfied, completing the proof of the theorem.

Equations (18) lend themselves to the calculation of explicit approximations to w and λ as power series in ε and q. Upon calculating these power series to terms of order two, one finds

$$\lambda = \gamma_1\varepsilon^2 + \gamma_2\varepsilon q + \gamma_3 q^2 + \text{higher order terms}, \tag{19}$$

where the coefficients γ_i can be written explicitly in terms of various combinations of the operators B, $D\,\frac{d^2}{dx^2}$, $L^{-1}Q$, P_1, $F(\cdot,0)$, and first order differentials of F, acting on the function φ. The fact that there are no first order terms in the expression for λ is a consequence of the facts that $F(\cdot,0)$ is quadratic or zero, that φ is a vector multiple of $\cos x$, and that $\frac{\partial\sigma_1}{\partial p}(0,p_0) = 0$ (in terms of Figure 3, $\sigma_1(\lambda_{cr},p)$ has a maximum at $p = p_0$).

If we further assume that $\frac{\partial^2\sigma_1}{\partial p^2}(0,p_0) < 0$ and $\frac{\partial\sigma_1}{\partial\lambda}(0,p_0) > 0$ (which inequalities are reasonable, from Assumption 1), then it follows that

$\gamma_3 > 0$. In this case, it can be seen that the "bifurcation diagram" is generically of one or the other of the two types shown in Figure 4. In that figure, the shaded regions consist of those pairs (λ, ε) for which a periodic solution $u = \varepsilon(\varphi + w)$ exists. These are local diagrams: ε, q, and λ are all assumed small.

In the context of problem (14), we have shown that when increasing the value of λ causes the trivial solution to lose its stability, it also causes many new periodic solutions to appear. There are two important questions now to be asked: (i) are there any other small bounded solutions besides the periodic ones we have constructed; and (ii) which of our new solutions, if any, are stable? These are difficult questions, and no complete answer to them exists at this time. However, some comments can be made.

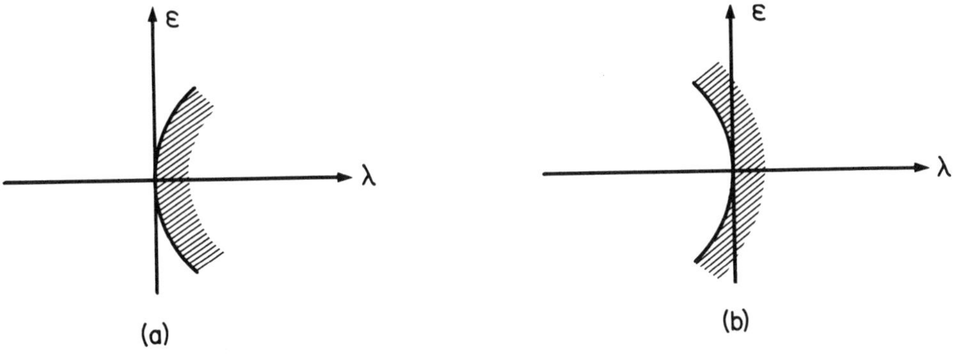

Figure 4.

The shaded areas are the totality of small (λ, ε) for which a periodic solution exists, with wave number near k_0 :
(a) the case $\gamma_4 \equiv \gamma_1 - \gamma_2^2(4\gamma_3)^{-1} > 0$; (b) the case $\gamma_4 < 0$.

Regarding the first question, we have constructed all possible periodic solutions with wave number near the one (k_0) with respect to which instability first sets in. This is clear by the implicit function theorem's uniqueness assertion. However, there probably exist quasiperiodic solutions as well (see Kirchgässner's treatment of another type of model problem in [17]). In fact, this is clearly true of (17) in the case $\varepsilon = 0$ (the limit equation for small amplitude solutions). This equation is linear with constant coefficients, and for each small $\lambda > 0$, there are exactly two values of q for which it has even periodic solutions. In terms of the original formulation (14), they correspond to two infinitesimally small even periodic solutions, with distinct periods. Such solutions of the linear problem may be superposed; linear combinations of them yield new quasiperiodic solutions (which are actually also periodic with possibly large period, if the two basic periods are rationally dependent). Each of the basic solutions may also be subjected to arbitrary translations in x, so they are no longer even. In this way, two more parameters are introduced, and in all we have a four-parameter family of quasiperiodic solutions of the linear problem. For fixed $\lambda > 0$, one expects quasiperiodic solutions to exist also for $\varepsilon > 0$, and there are ways to construct formal approximations to them in some cases. If it could be shown that a four parameter family of them exists, for some fixed $\lambda > 0$, then it would follow from a result on manifolds of neutral stability [8] that in fact there can be no other small bounded solutions besides those, and the answer to (i) would be more nearly complete.

As for the second question, concerning stability, no completely rigorous results will probably be obtained for some time. Nevertheless, heuristic results on linearized stability may be within reach. Specifically, on linearizing the operator on the left of (17) about a given periodic solution $u_{\varepsilon,q} = \varepsilon(\varphi + w_{\varepsilon,q})$, one obtains an operator with periodic coefficients, depending also on the parameters ε and q. One then would look for all eigenvalues of this operator, acting on C^2 functions, and ask for a relation between ε and q which will guarantee that no such eigenvalues lie in the right half-plane (0 is always an eigenvalue). Such pairs (ε,q) would then yield periodic solutions which may be stable in some sense.

By a Floquet-type argument, it is sufficient to pose this eigenvalue problem in the class of functions of the form $e^{i\alpha x}v(x)$ with v 2π-periodic and α real. Then the set of eigenvalues will depend on α as well as on ε and q, and one can restrict attention to small α, because otherwise all eigenvalues can be shown to have negative real part.

My conjecture is that in case (a) of Figure 4, only the solutions with extremal ε are stable in the above sense, whereas in case (b), none are stable except the trivial solution with $\lambda < 0$.

B. Larger amplitude patterns:

1. Peaks. There has been a sizeable amount of numerical simulation in recent years directed toward the discovery of finite amplitude patterns in model reaction-diffusion systems. The major efforts have been by a group in Brussels (see, for example, [3], [16]), and by Gierer and Meinhardt ([14], [19], [20]) in Tubingen. These efforts have indeed been successful, as one can judge by reading the cited papers.

The principal impetus has been biological; for example, the results bear on the formation of morphogenetic fields in developing organisms. The models of Gierer and Meinhardt have $n = 2$, the two reacting components being an "activator" u and an "inhibitor" v. These authors give arguments to explain why patterns may be expected, if the inhibitor diffuses more rapidly than the activator. (There is some analogy with Segel and Jackson's arguments involving stabilizers and destabilizers).

Here we deal with the following "inverse" problem: given a pair of functions $(\varphi(x),\psi(x))$, find a reaction-diffusion system with $n = 2$, for which (φ,ψ) is a pattern. Note that our theorem in Section 2 indicates that the corresponding problem with $n = 1$ does not have a solution, unless φ is monotone; and even then, the pattern is structurally unstable. We prove that for "single peak" distributions (φ,ψ) of a certain type, in which φ and ψ are related linearly, a solution of the inverse problem does exist (in fact, many solutions exist). We also show that for "multiple peak" distributions, solutions exist if the "stability" requirement for patterns is weakened to the statement that the spectrum of the linearized operator has no points in the unstable half-plane. It should be emphasized that it is easy to find systems for which a given distribution (φ,ψ) is a solution; but we must also prove that (φ,ψ) is stable for the system, a considerably more difficult task. The systems we construct are of the activator-inhibitor type, and the inhibitor does (as in [14], [19], [20]) diffuse more rapidly than the activator. Our results are not likely to be of practical importance, because model systems (1) which have been used in the past, and probably future models as well, have other desired properties besides merely being of activator-inhibitor type. Thus, the source functions f and g used in [14], [19], [20] were built on the basis of more or less specific morphogen reactions. At the same time, in constructing model reaction-diffusion systems with only two components, one should not attach overriding importance to having them mirror specific reaction networks involving the two species. In fact, the actual mechanisms modeled will involve a large number of reacting species; if one pictures the reduction to two as having been made through various pseudo-steady-state, slow reaction, or other approximations, then the connection between the source terms f and g and the actual kinetics will necessarily be obscured. In any case, the present analysis does shed light on the role which activation and inhibition mechanisms, with distinct migration rates, have in inducing stability of nonuniform structures.

We deal with a two-component system

$$u_t = u_{xx} + f(u,v) \tag{20a}$$

$$v_t = kv_{xx} + g(u,v) \tag{20b}$$

It has been nondimensionalized so that the diffusion coefficient of u is 1 ; we let k be that of v.

<u>Definition</u>: u is an <u>activator</u> if $f_u > 0$, $g_u > 0$, for the ranges of u and v at hand. v is an <u>inhibitor</u> if $f_v < 0$, $g_v < 0$ for this range.

Thus, increasing the amount of the activator increases the rate of production of u and v, whereas increasing v has the opposite effect.

First, we deal with "single peak distributions." Let $\varphi(x),\psi(x)$ be a pair of C^5 functions which, as in Figure 5, are even in x, approach limits as $|x| \to \infty$, satisfy

$$\varphi'(x) \neq 0 \quad \text{for} \quad x \neq 0, \ \varphi''(0) \neq 0, \ \varphi = h(\psi)$$

for some function h with $h'(v) > 0$ in the closure of the range of ψ, and such that $\dfrac{\varphi''(x)}{\varphi(x) - \varphi(\infty)}$ approaches a nonzero finite limit as $|x| \to \infty$.

<u>Theorem 1</u>. Let (φ,ψ) satisfy the above assumptions. Let $k > 1$. Then there exist functions $f(u,v)$, $g(u,v)$, and $p(v)$ such that $f_u > 0$, $g_u > 0$, $f_v < 0$, $g_v < 0$, $p > 0$, and (φ,ψ) is a C^0-stable stationary solution of

$$u_t = u_{xx} + f(u,v) \tag{21a}$$

$$v_t = k(p(v))^{-1}(p(v)v_x)_x + g(u,v) \ . \tag{21b}$$

Moreover if $\varphi = c_1\psi + c_2$ for some constants c_i with $c_1 > 0$, then $p \equiv 1$, so that our system is of the form (20).

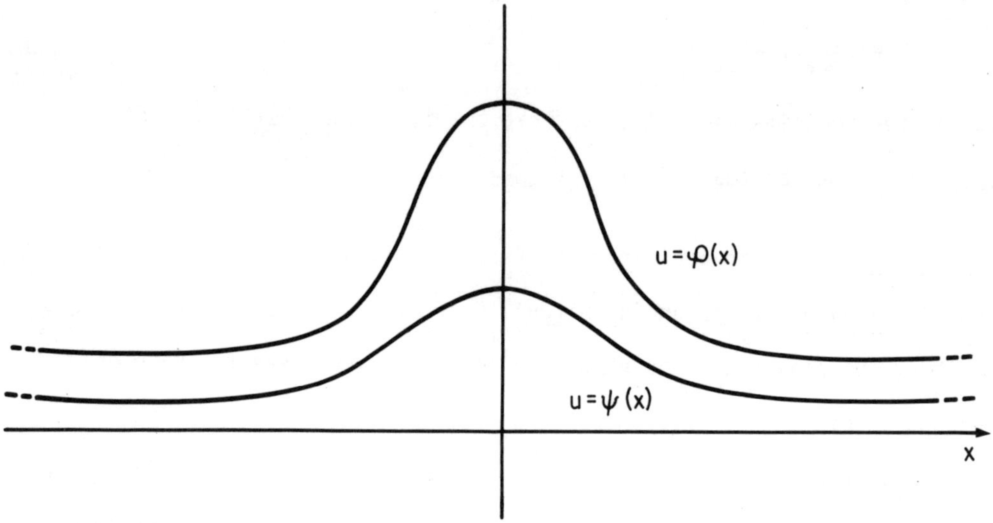

Figure 5
Typical single-peak distribution.

Remark: In fact, our construction will yield many solutions of the inverse problem.

Sketch of proof: First we consider the case $\varphi \equiv \psi$. Under the assumptions given, there is a C^2 function $F(u)$ such that $\varphi'' = -F(\varphi)$. The simplest system for which (φ,ψ) is a pattern will be of the form

$$u_t = u_{xx} + F(u) + \sigma(u - v) \tag{22a}$$

$$v_t = kv_{xx} + kF(v) + k\sigma(u - v) , \tag{22b}$$

where σ is a sufficiently large real number. Other examples will be shown later. It is clearly of activator-inhibitor type provided σ is large enough.

It is seen that (φ,ψ) is a solution of (22). We need to show it is stable. Let $\varepsilon = \frac{1}{k} < 1$, and let L be the linear operator $Lu \equiv u'' + F'(\varphi(x))u$. Let S be the linearization of the right hand side of (22) about (φ,ψ) :

$$S\begin{pmatrix}\bar{u}\\ \bar{v}\end{pmatrix} = \begin{pmatrix}L\bar{u} + \sigma(\bar{u} - \bar{v})\\ k(L\bar{v} + \sigma(\bar{u} - \bar{v}))\end{pmatrix},$$

as an operator from $C^2 \times C^2$ into $C^0 \times C^0$. For any given scalar λ, let $P_\lambda(L)$ be the operator

$$P_\lambda(L) \equiv \varepsilon\lambda^2 + \lambda[(1 - \varepsilon)\sigma - (1 + \varepsilon)L] + L^2 ,$$

acting on functions in $C^4(\underset{\sim}{R})$. If λ is such that $P_\lambda(L)$ is boundedly invertible on $C^0(\underset{\sim}{R})$, then it turns out that $(S - \lambda)$ is boundedly invertible on $C^0(\underset{\sim}{R}) \times C^0(\underset{\sim}{R})$, so that $\lambda \notin \Sigma(S)$ (spectrum of S). This is proved by actually writing down a solution of $(S - \lambda)\begin{pmatrix}\bar{u}\\ \bar{v}\end{pmatrix} = \begin{pmatrix}p\\ q\end{pmatrix}$ in terms of $P_\lambda(L)^{-1}$ and L. Hence $\Sigma(S) \subset \{\lambda : \Sigma(P_\lambda(L)) \ni 0\} \equiv \Lambda$, so that information about the spectrum of S can be found from knowledge of $\Sigma(P_\lambda(L)) = P_\lambda(\Sigma(L))$.

The following facts about the function $P_\lambda(\mu)$ can be verified directly: For fixed μ, it is a quadratic polynomial in λ. Given any M, its two roots will be real and negative for any $\mu < M$ and $\mu \neq 0$, as long as $0 < \varepsilon < 1$ and $(1 - \varepsilon)\sigma$ is large enough (depending on M). (Incidentally, this is where the condition $k > 1$ is seen to be necessary.) If $\mu = 0$, however, one of the roots will be zero and the other negative. Now the spectrum $\Sigma(L)$ is, in fact, real, bounded from above by some number M, and contains 0 as a simple isolated eigenvalue. This follows from our assumptions on φ, and from known facts about the spectra of second order operators. Thus as μ ranges over $\Sigma(L)$, the set of roots λ of $P_\lambda(\mu)$ will range over a set of nonpositive real numbers which contains the origin once, and is otherwise bounded away from zero. In view of the fact (shown in the previous paragraph) that $\Sigma(S) \subset \Lambda$, we know that $\Sigma(S)$ has the same property.

But with this knowledge about the spectrum of S, we can apply Sattinger's stability theorem [24] to deduce the C^0-stability of the solution $u = \varphi$, $v = \varphi$. (Incidentally, I believe this to be one of the first examples in which Sattinger's theory has been applied to specific systems of order greater than one.)

Now consider the case when $\varphi = h(\psi) \not\equiv \psi$. We write the system (22) with v replaced by the symbol w, then transform the second dependent variable according to the equation $w = h(v)$, to obtain a system of the form (21), with $p(v) = h'(v)$. Clearly if $h' = \text{const}$, we may take $p \equiv 1$. This completes the proof.

Next, we consider distributions with many peaks, in fact those which are periodic in x. Let $\varphi(x)$ and $\psi(x)$ be periodic C^5 functions such that $\varphi'' = -F(\varphi)$ for some C^2 function F, and $\varphi = c_1\psi + c_2$, $c_1 > 0$.

Theorem 2. Let φ and ψ satisfy these assumptions. Let $k > 1$. Then there exist functions f, g satisfying the inequalities stated in Theorem 1, such that (φ,ψ) is a solution of (20). Moreover, the spectrum of S (the linearization of the right hand side of (20) about $u = \varphi$, $v = \psi$) is contained on the nonpositive real axis.

Remark: The fact that none of the spectrum of S is in the unstable half-plane is an indication of the stabilizing influence of the activator-inhibitor mechanism, although it is not proved that (φ,ψ) is stable in the C^0 sense.

The proof of Theorem 2 proceeds as did the other theorem, with the difference that it is no longer true that 0 will be an isolated point of the spectrum of L.

<u>Other solutions of the inverse problem</u>: Our solutions of the inverse problem (in the case $\varphi = \psi$) have been of the form (22). Equally valid solutions can be obtained by replacing the constant σ in the second equation (22b) by a different large constant τ in a certain range depending on σ, and by adjoining extra terms $(u - v)^2\tilde{f}(u,v)$ and $(u - v)^2\tilde{g}(u,v)$ to the right sides of (22a) and (22b), respectively, where $\tilde{f}$ and $\tilde{g}$ are arbitrary. The given distributions will of course still be solutions of the revised equations, and the stability analysis is unchanged. Moreover, the inequalities given in the statement of the theorem hold true for u near φ, v near ψ, but are no longer necessarily true for arbitrary u and v.

<u>Example</u>: The function $\varphi(x) = \dfrac{6e^{x/d}}{(1 + e^{x/d})^2}$ is an even peak-like distribution with $\varphi(0) = 3/2$, $\varphi(\infty) = 0$, and with peak-width of the order d. It can be checked that φ satisfies the equation $\varphi'' + d^{-2}(\varphi^2 - \varphi) = 0$. Setting $\psi(x) = a\varphi(x)$, $a > 0$, we see that (φ,ψ) will be an activator-inhibitor pattern for the system

$$u_t = u_{xx} + d^{-2}(u^2 - u) + \sigma(u - av) + (u - av)^2\tilde{f}(u,v) \quad ,$$

$$v_t = kv_{xx} + kd^{-2}(av^2 - v) + \sigma(u - av) + (u - av)^2\tilde{g}(u,v) \quad ,$$

provided that $k > 1$, and σ is sufficiently large. The functions $\tilde{f}$ and $\tilde{g}$ are arbitrary.

2. Plateaus. The previous section is relevant when one wishes to model the concentration of reacting substances at isolated locations. The present section, on the other hand, is concerned with modeling phenomena of differentiation, characterized by the development of sharp boundaries between regions within which the concentrations are relatively uniform. We call these sharp boundaries "transition layers" (see [9]).

The methods of singular perturbations may be used to construct patterns with sharp transition layers. This was explained in [10] for boundary value problems on finite domains in higher dimensional space, and an example was given in [9] of transition layer solutions in one independent variable, though no discussion of stability was given in the latter paper.

Here we review the construction process for periodic plateau-type patterns on the whole line. In this case, contrary to that studied in [10], there will be no boundary effect on the appearance of our dissipative structures, because there is no boundary. The following will be formal and nonrigorous.

As in [10], we take as model system the following:

$$u_t = \varepsilon^2 u_{xx} + f(u,v) \tag{23a}$$

$$v_t = v_{xx} + g(u,v) \ , \tag{23b}$$

where ε is small, and where the source term f has the characteristics shown in Figure 6 (see [11] for references to reaction networks realizing this type of source function).

Our object will be to construct periodic stationary solutions having the appearance shown in Figure 7. As seen there, the upward transitions occur roughly at locations α_n with $\alpha_n = \alpha_0 + np$, where p is the wave-length; and the downward transitions are at $\beta_n = \beta_0 + np$. On the relatively "flat" portions between transitions, (23a) with $u_t = \varepsilon = 0$ should be satisfied

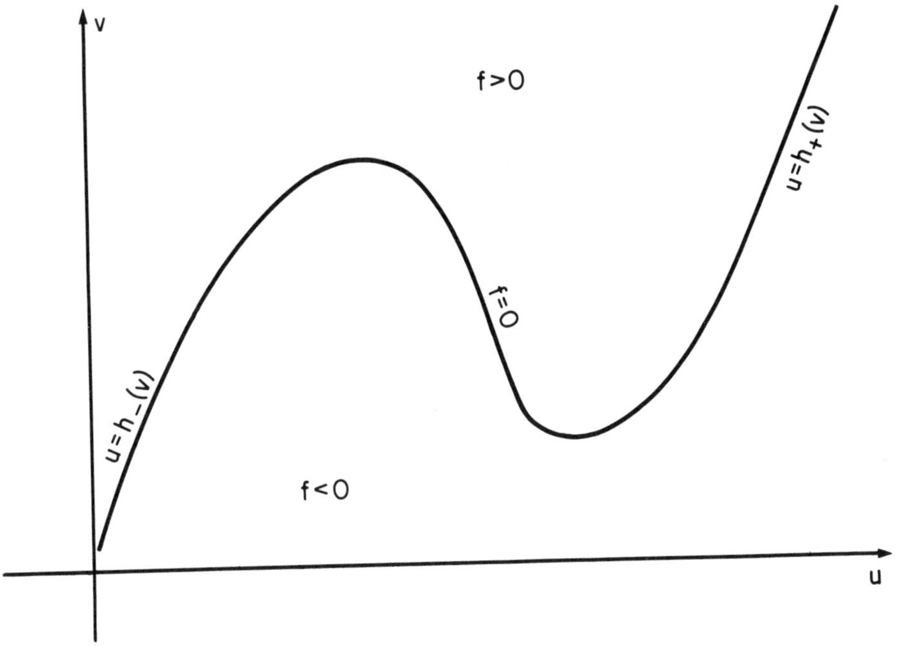

Figure 6.
Regions of positivity and negativity for
the source function $f(u,v)$.

approximately. That is, $f(u,v) = 0$, so the phase plane image should lie near the S-shaped curve in Figure 6. Referring to that figure, where the two ascending branches $u = h_{\pm}(v)$ are attractors for the corresponding kinetic equations, we specify that the lower flat regions $\beta_n < x < \alpha_{n+1}$ be mapped onto the branch $u = h_-(v)$, and the upper portions $\alpha_n < x < \beta_n$ be mapped onto $u = h_+(v)$.

We are seeking solutions for which v does not change abruptly through the transition layer, although u does. Therefore to lowest order, there is a well-defined value of v at each transition point. It is explained in [10] that the only value of v which can support a transition layer is a value v^* for which

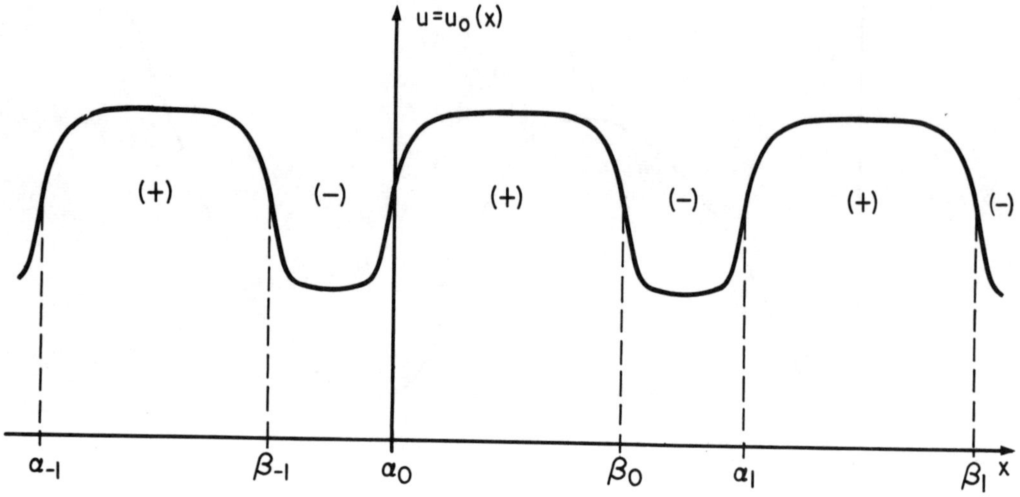

Figure 7.
Plateau-type pattern for the u-component.

$$J(v^*) = 0 , \tag{24}$$

where $J(v) \equiv \int_{h_-(v)}^{h_+(v)} f(u,v)du$. Let us assume there is only one value of v satisfying (24). The explanation of (24), as well as the construction of the solution inside the layer itself, is along the following lines. Suppose the transition occurs at $x = 0$, where $v = v_0$. We stretch variables by setting $\xi = x/\varepsilon$, then (23a) with $u_t = 0$ becomes, to lowest order in the new variables,

$$\frac{d^2u}{d\xi^2} + f(u,v_0) = 0 . \tag{25}$$

The methods of matched asymptotics dictate that (25) should be solved for $\xi \in (-\infty,\infty)$ with boundary conditions (if it is an upward transition) $u(-\infty) = h_-(v_0)$, $u(+\infty) = h_+(v_0)$. Assuming a solution $u(\xi)$ exists, multiply

(25) by $\frac{du}{d\xi}$ and integrate the result from $-\infty$ to $+\infty$ to obtain

$$0 = \int_{-\infty}^{\infty} [\frac{1}{2} \frac{d}{d\xi}((\frac{du}{d\xi})^2) + f(u,v_0)\frac{du}{d\xi}]d\xi = \int_{h_-(v_0)}^{h_+(v_0)} f(u,v_0)du = J(v_0) ;$$

hence $v_0 = v^*$. It turns out that this is a sufficient, as well as a necessary, condition for the existence of a solution of (25) with the given boundary conditions.

At this point, we have obtained the following information. On the broad tops and bases of the plateau configurations in Figure 7, u and v are related by $u = h_{\pm}(v)$. At the transition points between them, $v = v^*$. We now proceed to a determination of $v(x)$. In (23b), we replace the variable u by $h_{\pm}(v)$, depending on whether we are on the top or the base of a plateau. We thus ignore the fine structure of the transition, replacing it by an actual discontinuity. Then (23b) (with $v_t = 0$) becomes

$$v_{xx} + G(v,x) = 0 , \tag{26}$$

$$\text{where} \quad G(v,x) = \begin{cases} g_-(v) \equiv g(h_-(v),v) & \text{for } \beta_n < x < \alpha_{n+1} , \\ g_+(v) \equiv g(h_+(v),v) & \text{for } \alpha_n < x < \beta_n . \end{cases}$$

This equation, together with the condition $v = v^*$ for $x = \alpha_n$ or β_n, can be expected typically either to have no solution, or a one-parameter family of periodic solutions (translations in x would give a second parameter, which we ignore). For example, suppose $g_-(v) < 0$. Then $v_{xx} > 0$ for $x \in (\beta_n,\alpha_{n+1})$, which implies $v < v^*$ for $x \in (\beta_n,\alpha_{n+1})$ and $v_x(\beta_n) < 0$, $v_x(\alpha_n) > 0$. If $g_+(v) < 0$ as well, then the same derivative conditions must hold at the endpoints of the intervals (α_n,β_n). But these two sets of derivative conditions contradict one another, so there can be no solution.

For a pattern to exist, it is therefore necessary that $g_{\pm}(v)$ not be of one sign. As a simple example, let us assume that $g_{\pm}(v)$ are constants with $g_- < 0 < g_+$. Then from (26), we have $v'' = -g_+$ for $x \in (\alpha_n, \beta_n)$; hence setting $p_+ = \beta_n - \alpha_n$, we get

$$v = v^* + \frac{1}{2} g_+(x - \alpha_n)(\beta_n - x),\ x \in (\alpha_n, \beta_n);\ v_n'(\beta_n) = -\frac{1}{2} g_+ p_+ .$$

Similarly, $v = v^* + \frac{1}{2} g_-(x - \beta_n)(\alpha_{n+1} - x)$ in (β_n, α_{n+1}), $v_n'(\beta_n) = \frac{1}{2} g_- p_-$, where $p_- = \alpha_{n+1} - \beta_n$ (so that $p_- + p_+ = p$). Matching the derivatives at β_n yields the condition

$$g_+ p_+ = -g_- p_- \tag{27}$$

Thus in this simple example, we apparently obtain a plateau-like solution for any spacing $p_{\pm}$ of the α's and β's satisfying (27). However, this is only partly true: the wavelength p cannot be too large, for then the range of the function v might extend beyond the domains of one or both of the functions $h_{\pm}$.

The foregoing describes a method of obtaining formally approximate solutions for small ε. The procedure could be continued to yield higher order approximations; the constructions in [10] and [11] are relevant here.

Now supposing a plateau solution (which we denote by $(u_0(x), v_0(x))$) has been constructed, how can we infer its stability or instability? The following heuristic argument suggests a workable criterion.

Suppose we subject the given stationary solution to a perturbation which is small in the C^0 norm. Such a perturbation, of course, does not erase or relocate the transition layers. For the moment, let us assume that the latter do not, in fact, move during the subsequent evolution of the perturbed system. Then between the layers, the image points (u,v) will again be drawn to the attractors $h_{\pm}(v)$ (by neglecting ε in (23a)), and at the same time, the evolution of v will be governed by (23b). Eventually, the term g in this latter equation may be replaced by $G(v,x)$ (see (26)):

$$v_t = v_{xx} + G(v,x) \quad . \tag{28}$$

The solution before perturbation was such that v_0 satisfied (28) and was time-independent, of course. So we must now investigate the stability of solutions of (28). If we impose the condition

$$\frac{d}{dv} g_{\pm}(v) < 0 \tag{29}$$

(this being the v-derivative along the ascending branches in Figure 6), then the solution $v_0(x)$ will be stable. In fact, for small positive constants δ, $v_0(x) \pm \delta$ will be super- and sub-solutions of (28), so again as in [1], solutions of (28) with initial data caught between $v_0 - \delta$ and $v_0 + \delta$ will converge uniformly to v_0.

This argument asserts stability under condition (29) if we assume the transition layers are immobile. They are not, in fact, immobile, but their movement is slow and can be analyzed by the techniques in [11]. Immediately following the initial perturbation, the value of v at any given transition layer will no longer necessarily satisfy $J(v) = 0$ (as in (24)). What this means is that the layer will develop into a sharp wave front which, locally, will be of the form $u = U(\frac{x - \varepsilon ct}{\varepsilon}, x)$, where c depends on v at the position of the front, so will itself change with time. This approximate representation of u follows the argument in [10], [11], to which the reader is referred for more details. As stated above, the velocity εc will be small, so that until a time of the order $1/\varepsilon$ has elapsed, the above assumption that the layer is immobile should be valid.

But after this time, shifts in the transition layer's position due to the slow motion of the front may have accrued.

The effect of changes in the layer's position should therefore be determined. It turns out that the nature of this effect depends on the signs of the two functions $g_{\pm}(v)$. In the following, we call the collection of intervals on which $u \sim h_+(v)$ the domain of the $(+)$ state, and similarly for the $(-)$ state. Suppose, for example, that $g_-(v) < 0 < g_+(v)$. It will then also be true that $g(u,v) < 0$ when the point (u,v) is in a neighborhood of the left ascending branch in Figure 6, and $g > 0$ near the right ascending branch. We focus attention on the effect of shifts in the position of a single transition layer. Suppose, initially, that we are in a stationary state, and that the points (u,v) lie in the two neighborhoods described above on opposite sides of the layer. Now suppose the position of the layer is perturbed in the direction which will increase the "domain" of the state $(+)$ (that is to say, it is moved away from the $(+)$ state toward the $(-)$ state. Since $g_+ > g_-$, the immediate effect of this is to increase the source term $g(u,v)$ in (23b), in the interval between the old and the new positions. This will cause the function $v_t(x,t)$, hence v, to increase. At the same time, the value of v at the layer has also changed by virtue of the fact that the initial stationary function $v(x)$ was not constant, whereas the layer's position was altered. This second effect may, for a period of time, counteract the first effect (the increase of v due to increasing the source terms); but eventually the first will dominate, because the subsequent motion of the layer is slow, and v_t will remain positive until the position has moved a significant amount. So eventually the value of v at the layer will surpass its initial value.

Let us now assume that $J(v)$ is an increasing function of v (apparently the case for the function f depicted in Figure 6). Then such an increase in the value of v at the front also raises J, which in turn results in the (+) state becoming more dominant, in the terminology of [10] and [11]. This means that the (slow) motion of the front will proceed in a direction which further increases the domain of the (+) state. The perturbed layer will therefore not tend to return to its original location, and we are in an unstable situation.

On the other hand, if we assume the opposite inequalities

$$g_+(v) < 0 < g_-(v) \ , \tag{30}$$

then the above analysis indicates that a perturbed position of the layer will eventually move back toward its original location. Essentially the same analysis holds if the perturbation is in the opposite direction.

On the basis of the above arguments, which admittedly are far from being rigorous, I conjecture that the constructed periodic plateau-type stationary solution will be stable if (29) and (30) hold, and if J is an increasing function of v.

In [10], an argument for stability or "realizability" was given, in some respects similar to the above. Yet in other respects, it was different, because the effects of boundary conditions in a bounded domain were essentially involved.

4. DISCUSSION

This paper is devoted entirely to patterns on the entire real line, and it may well be objected that real reacting and diffusing systems are always finite. Suppose the real system is one-dimensional, and its domain is a finite interval I. If the length of I is large compared with the characteristic length of the patterns formed, then the infinite-line approximation should be a good one, in the interior of I. In fact, the patterns studied here are generated by the processes of reaction and diffusion only, and so should exist in interior regions independently of any boundary effects. Of course there also exist patterns which are generated or strongly conditioned by the presence of boundaries and by the transport mechanisms occurring there. This type of pattern is not the subject of the paper. In short, our restriction to the infinite line was made for the purpose of exploring pattern phenomena generated solely by reaction and diffusion, without having to worry about boundary effects.

Nevertheless, some things can be said immediately about certain patterns on finite domains, analogous to those discussed here.

Consider first the scalar case, $n = 1$. This is a matter of solving (2) on an interval, with given boundary conditions. We restrict attention to homogeneous Neumann (no-flux) or to Dirichlet $(u = 0)$ boundary conditions. Again, all solutions can be constructed, and their stability analyzed by techniques similar to those described. It turns out that under Neumann conditions, the only stable solutions are certain constant ones, and under Dirichlet conditions, the only stable ones are those which have a single internal extremum (except in the case when $u \equiv 0$ is stable). See also the treatment in [5].

Consider next the question of small amplitude patterns, as treated in 3A. Set on a finite interval with homogeneous boundary conditions, this is a standard bifurcation problem, and the results of [6], [7], [23] apply. One commonly finds stable super-critical bifurcation branches of nonconstant solutions which, for small amplitudes, approximate periodic functions with many wavelengths fitted into the given interval ([2], [4]). These will also approximate some of the periodic solutions found in 3A .

The treatment of peaks--single or multiple--in 3B.1 may be extended with little change to the case of a finite interval with (say) Neumann boundary conditions imposed. (Of course, one must begin with distributions φ and ψ which, themselves, satisfy these boundary conditions). The same is true of the plateaus in 3B.2 . See especially [9] and [10].

A second objection may be that real reacting and diffusing systems are distributed in space or on a surface, rather than on a line, as we have here. It is probable that the extension of the present results to problems on $\underset{\sim}{R}^2$ or $\underset{\sim}{R}^3$ will entail many complications and difficulties. In the case $n = 1$, for example, one can no longer immediately write down all stationary solutions of $\Delta u + f(u) = 0$; and though in principle the technique of Aronson and Weinberger works in space, its application may be difficult.

As for small amplitude patterns in the case $n > 1$, this problem is fraught with most of the difficulties associated with the Benard problem in fluid mechanics. One can find solutions which are doubly periodic with various kinds of symmetries, but the stability question is very difficult, unless one allows only perturbations within certain classes restricted by artificial symmetry conditions. Of course, if one wants to find patterns for which boundary effects are important, one has a boundary value problem in a bounded domain. Typically, this reduces to a standard bifurcation problem, and the identification of the bifurcation points, together with the determination of the approximate shape of the small amplitude patterns which arise, involves solving a linear eigenvalue problem.

There are real complications involved in extending the treatment in 3B.1 to higher dimensions. On the other hand, plateau configurations (as in 3B.2) can be constructed, and was done in [10] (see also [12]).

A final comment should be made about the terminology "pattern formation." The same term is also commonly used in developmental biology, where it refers to structural organization in developing organisms. The possible relevance of reaction-diffusion patterns, such as those in the present paper, to patterns in this other sense, will suggest itself. No claim is being made here that reaction-diffusion patterns are anywhere near to being adequate models for the biological patterns. In fact, the latter are undoubtedly highly complicated phenomena involving many other types of processes besides those suggested here. At the same time, any serious student of theoretical developmental biology should be aware of the rich repertoire of solutions, including stable patterned ones, exhibited by rather simple models based on reaction and diffusion alone. To this extent, I hope the results given in this paper may be useful to the biologist.

REFERENCES.

1. D. G. Aronson and H. F. Weinberger, Nonlinear diffusion in population genetics, combustion, and nerve propogation, in: Proceedings of the Tulane Program in Partial Differential Equations and Related Topics, Lecture Notes in Mathematics, No. 446, ed. J. Goldstein, Springer, Berlin, 1975.

2. J. F. G. Auchmuty and G. Nicolis, Bifurcation analysis of nonlinear reaction-diffusion equations. I. Evolution equations and the steady state solutions, Bull. Math. Biol. 37 (1975), 323.

3. A. Babloyantz and J. Hiernaux, Models for cell differentiation and generation of polarity in diffusion-governed morphogenetic fields, Bull. Math. Biol. (to appear).

4. J. A. Boa and D. S. Cohen, Bifurcation of localized disturbances in a model biochemical reaction, SIAM J. Appl. Math. 30 (1976), 123-135.

5. N. Chafee and E. Infante, A bifurcation problem for a nonlinear parabolic equation, J. Diff. Eq. 15 (1974), 17-37.

6. M. G. Crandall and P. H. Rabinowitz, Bifurcation from simple eigenvalues, J. Functional Anal. 8 (1971) 321-340.

7. M. G. Crandall and P. H. Rabinowitz, Bifurcation, perturbation of simple eigenvalues, and linearized stability, Arch. Rat. Mech. Anal. 52 (1973) 161-180.

8. C. Conley and N. Fenichel, private communications.

9. P. Fife, Boundary and interior transition layer phenomena for pairs of second-order differential equations, J. Math. Anal. Appl. 54 (1976) 497-521.

10. P. Fife, Pattern formation in reacting and diffusing systems, J. Chem. Phys. 64 (1976) 554-564.

11. P. Fife, Singular perturbation and wave front techniques in reaction-diffusion problems, Proc. AMS-SIAM Symposium on Asymptotic Methods and Singular Perturbations, New York, 1976, 23-50.

12. P. Fife and W. M. Greenlee, Transition layers for elliptic boundary value problem with small parameters, Uspehi Mat. Nauk, 24 (1974) 103-130; Russ. Math. Surveys 29 (1975) 103-131.

13. P. Fife and J. B. McLeod, The approach of solutions of nonlinear diffusion equations to travelling front solutions, Arch. Rat. Mech. Anal. to appear. Announcement in Bull. Amer. Math. Soc. 81 (1975) 1075-1078.

14. A. Gierer and H. Meinhardt, A theory of biological pattern formation, Kybernetik 12 (1972) 30-39.

15. D. I. Gmitro and L. E. Scriven, A physicochemical basis for pattern and rhythm, in: Intracellular Transport, ed. K. B. Warren, Academic Press, New York (1966) 221-255.

16. M. Herschkowitz-Kaufman and G. Nicolis, Localized spatial structure and nonlinear chemical waves in dissipative systems, J. Chem. Phys. 56 (1972) 1890-1895.

17. K. Kirchgässner, Pattern selection and cellular bifurcation in fluid dynamics, in: Applications of Bifurcation Theory, Proceedings of Advanced Seminar on Applications of Bifurcation Theory, Academic Press, 1977 (to appear).

18. S. A. Levin, Spatial patterning and the structure of ecological communities, in: Lectures on Mathematics in the Life Sciences, Vol. 8, Amer. Math. Soc., Providence (1976) 1-35.

19. H. Meinhardt, The formation of morphogenetic gradients and fields, Ber. Deutsch, Bot. Ges. 87 (1974) 101-108.

20. H. Meinhardt and A. Gierer, Applications of a theory of biological pattern formation based on lateral inhibition, J. Cell. Sci. 15 (1974) 321-346.

21. H. G. Othmer, Current problems in pattern formation, in: Lectures on Mathematics in the Life Sciences, Vol. 9, S. A. Levin, ed., Amer. Math. Soc., Providence, (1977).

22. I. Prigogine and G. Nicolis, On symmetry-breaking instabilities in dissipative systems, J. Chem. Phys. 46 (1967) 3542-3550.

23. D. H. Sattinger, Topics in Stability and Bifurcation Theory, Springer-Verlag, Lecture Notes #309 (1973).

24. D. H. Sattinger, On the stability of waves of nonlinear parabolic systems, Advances in Math. (to appear).

25. L. A. Segel and J. L. Jackson, Dissipative structure: an explanation and an ecological example, J. Theor. Biol. 37 (1972) 545-559.

26. L. A. Segel and S. A. Levin, Application of nonlinear stability theory to the study of the effects of dispersion on predator-prey interactions, in: Amer. Inst. Physics Symp., Vol. 27 (1976).

27. A. M. Turing, The chemical basis for morphogenesis, Phil. Trans. Roy. Soc. (London) B237 (1952) 36.

Paul C. Fife
Department of Mathematics
University of Arizona
Tucson, Arizona 85721

D HENRY

Gradient flows defined by parabolic equations

One of the simplest kinds of ordinary differential equations is the gradient flow: given a smooth real-valued function V, always go downhill.

$$u_t = -\text{grad } V(u)$$

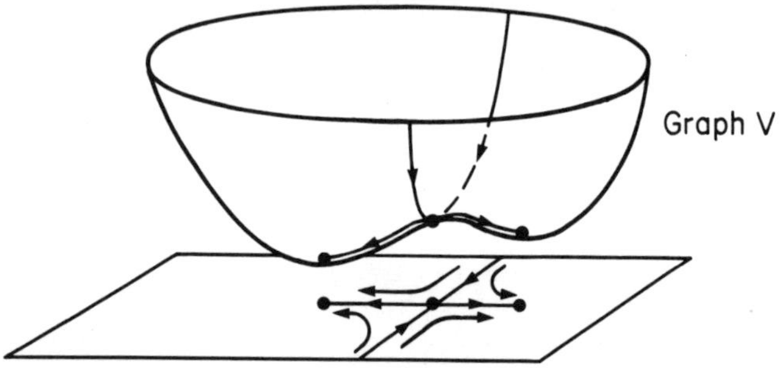

We consider an analogous class of problems in an infinite dimensional state space. The proofs are merely sketched; for the details see [1].

In a bounded smooth domain $\Omega \subset R^n$ consider the initial boundary-value problem

$$u_t = \sum_{i,j=1}^{n} (a_{ij}(x)u_{x_i})_{x_j} + f(x,u) \quad \text{on} \quad \Omega \times R^+ \qquad (*)$$

$$u = 0 \quad \text{on} \quad \partial\Omega \times R^+ \quad \left[\text{or } \frac{\partial u}{\partial \nu} + h(x)u = 0 \quad \text{on} \quad \partial\Omega \times R^+ \right]$$

$$\nu = \text{outward conormal}$$

$$u(x,0) = u_0(x), \quad x \in \Omega .$$

We make the following assumptions [and items in square brackets refer to the second boundary condition].

(1) Smoothness: a_{ij}, f, $\frac{\partial f}{\partial u}$ [and h] are continuously differentiable; $\{a_{ij}(x)\}_{i,j=1}^n$ is uniformly positive definite [and $h(x) \geq 0$].

(2) Coerciveness: $\limsup_{u \to \pm\infty} \frac{f(x,u)}{u} \leq 0$ uniformly in x [and it is strictly negative if $h(x) \equiv 0$].

(3) Polynomial growth: If $n > 1$, $|f(x,u)| \leq b + c|u|^q$ on $\Omega \times R$ for some constants b, c, q.

Remark. Assumption (2) ensures that all solutions are bounded: the linear terms pull them in from infinity where the nonlinear terms create no problem. Assumption (3) may not be necessary, but is an improvement over Auchmuty's assumption [2] that $q < (n + 2)/(n - 2)$.

We work in the state-space $X = W_0^{1,p}(\Omega)$ [or $W^{1,p}(\Omega)$] where $2 \leq p < \infty$, $p > n$, so $X \subset C(\Omega)$. We call (*) a gradient flow since it is formally

$$u_t = -\text{grad } V(u)$$

where

$$V(u) = \int_\Omega \{\tfrac{1}{2} \Sigma a_{ij}(x) u_{x_i} u_{x_j} - F(x,u)\}dx + [\int_{\partial\Omega} \tfrac{1}{2} h(x)u^2 ds] , \quad u \in X ,$$

$$F(x,u) = \int_0^u f(x,s)ds.$$

It is easily proved that, if $u(x,t)$ is a solution of (*) ,

$$\frac{d}{dt} V(u(\cdot,t)) = - \int_\Omega u_t^2 dx$$

so the solution always moves downhill except at an equilibrium point.

Theorem 1. Let $u_0 \in X$. Then there is a unique solution u of (*) which remains bounded in X as $t \to +\infty$, and in fact remains in a compact set in X. The set of limit points of this solution $\omega(u_0)$ is a nonempty compact connected subset of E, the set of equilibria

$$E = \{\varphi \in X \mid \text{grad } V(\varphi) = 0\},$$

and $u(\cdot,t) \to \omega(u_0)$ in X as $t \to +\infty$.

The key to the proof of Theorem 1 is establishment of an a priori bound on $||u(\cdot,t)||_{W^{1,p}(\Omega)}$, which follows from bounds on $||f(\cdot,u(\cdot,t))||_{L^p}$, which follows from bounds on $||u(\cdot,t)||_{L^k(\Omega)}$ with $k = pq$, when $n > 1$. If $\dim \Omega = 1$, the result follows more easily from bounds on u in $W^{1,2}(\Omega) \subset L^\infty(\Omega)$, since V bounds the square of the $W^{1,2}(\Omega)$-norm.

Theorem 2. Let $\varphi \in E$ be a simple solution, i.e. hyperbolic, i.e. φ is an equilibrium and the linearization of $-\text{grad } V$ about φ,

$$\sum_{i,j=1}^{n} \frac{\partial}{\partial x_i} a_{ij} \frac{\partial}{\partial x_j} + \frac{\partial f}{\partial u}(x,\varphi(x))$$

plus boundary conditions, is invertible.

Then there exist C^1 invariant submanifolds of X, $W^s(\varphi)$ and $W^u(\varphi)$,

$$W^s(\varphi) = \{u_0 \in X \mid u(\cdot,t) \to \varphi \text{ as } t \to +\infty\}$$

$$W^u(\varphi) = \{u_0 \in X \mid \text{there is a solution } u \text{ of } (*) \text{ on } -\infty < t < 0 \text{ with } u|_{t=0} = u_0 \text{ and } u(\cdot,t) \to \varphi \text{ as } t \to -\infty\}$$

with $\dim W^u(\varphi) = \text{codim } W^s(\varphi) =$ number of positive eigenvalues of the linearization about φ, which is finite.

The proof of this starts with an integral equation analogous to that used for the ordinary differential equation case, which gives the local stable and unstable manifolds. These are extended globally by following the solutions of $(*)$, using the implicit function theorem and backward uniqueness for parabolic equations (see [3]). That these are embedded submanifolds, not merely immersed, follows (roughly) from the fact that a gradient flow cannot bend back on itself (see [1]).

Corollary. Suppose E is simple, i.e. every equilibrium is simple: this says V is a Morse function. Then

(i) E is finite;

(ii) $X = \bigcup_{\varphi\in E} W^s(\varphi)$, a disjoint union of open sets - domains of attraction of the stable equilibria - together with a closed nowhere dense remainder of codimension one;

(iii) $K = \bigcup_{\varphi\in E} W^u(\varphi)$ is the maximal bounded invariant set, a compact connected finite-dimensional set in X with

$$K = \bigcap_{t\geq 0} \overline{u(t;B)}$$

where B is any sufficiently large ball about 0 in X and $u(t;B)$ is its image under the flow defined by $(*)$ after time t.

Since E is simple, it is discrete; and E is easily proved compact, hence it is finite. For each $u_0 \in X$ the solution $u(\cdot,t) \to \varphi$ for some $\varphi \in E$, by Th. 1 and the discreteness of E, so (ii) follows. If B is any ball in X, $u(t;B)$ is precompact for each $t > 0$, and it is easily proved that there exists a maximal bounded invariant set K and it is compact and connected. If $u_0 \in K$, there exists a bounded solution through u_0 of $(*)$ on $-\infty < t < \infty$; it follows that the solution tends to a point of E as $t \to -\infty$, so $u_0 \in W^u(\varphi)$ for some φ and (iii) follows.

The assumption that E is simple (i.e. V is a Morse function) is true for most choices of $\{a_{ij}\}$, f, h and Ω, separately or together, in the sense of Baire category (see [1]). Unfortunately this doesn't help us in any particular problem. We describe two examples where the hypothesis can be verified.

Example 1. (Chafee and Infante [4]).

$$u_t = u_{xx} + \lambda f(u) \ , \ 0 < x < \pi$$

$$u = 0 \ \text{at} \ x = 0, \ x = \pi$$

where $\lambda \geq 0$, f is C^2 with $f(0) = 0$ $f'(0) = 1$, $uf''(u) < 0$ for $u \neq 0$, and $\limsup_{u \to \pm\infty} f(u)/u \leq 0$.

It has been proved [4] that for $n^2 < \lambda < (n + 1)^2$ for an integer $n \geq 0$, E is simple and consists of $2n + 1$ points $0, \varphi_1^{\pm}, \ldots, \varphi_n^{\pm}$ where $\varphi_k^{\pm}(x)$ vanishes $(k - 1)$ times in $0 < x < 1$ with

$$\pm \frac{d}{dx} \varphi_k^{\pm}(0) > 0 \ .$$

These satisfy

$$\dim W^u(0) = n$$

$$\dim W^u(\varphi_k^{\pm}) = k - 1 \ \text{for} \ k = 1, \ldots n$$

with $V(0) > V(\varphi_n^{\pm}) > \ldots > V(\varphi_2^{\pm}) > V(\varphi_1^{\pm})$.

If $0 < \lambda < 1$, the only equilibrium is $u = 0$ and it is stable: $W^s(0) = H_0^1(0,\pi)$. If $\lambda > 1$, the equilibria $\varphi_1^{\pm}$ are stable and all other equilibria are unstable, so from the Corollary we have

$$H_0^1(0,\pi) = W^s(\varphi_1^+) \cup W^s(\varphi_1^-) \cup \{\text{a closed nowhere dense set}\} \ .$$

When f is odd (for example, $f(u) = u - u^3$) and $0 < \lambda < 9$, we can display the maximal bounded invariant set K :

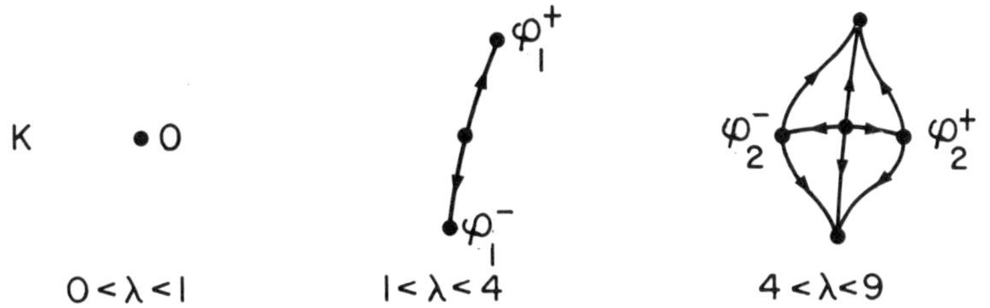

If $n^2 < \lambda < (n+1)^2$, $\dim K = \dim W^u(0) = n$, but the details of the flow in K are not known for $n \geq 3$ $(\lambda > 9)$.

Example 2. (Fisher, Fleming [5], Henry [6])

$$u_t = \Delta u + \lambda s(x) f(u) \quad \text{on} \quad \Omega \times R^+$$

$$\frac{\partial u}{\partial n} = 0 \quad \text{on} \quad \partial\Omega$$

where f is C^2, $f(0) = f(1) = 0$, $f(u) > 0$ on $0 < u < 1$, $s : \Omega \to R$ is C^1 and $\lambda \geq 0$. It is further assumed that Ω is a bounded smooth domain and $f''(u) < 0$ for $0 \leq u \leq 1$.

If $0 \leq u_0 \leq 1$ a.e. then $0 \leq u \leq 1$ on $\Omega \times R^+$ by the maximum principle, and we need not consider behavior of f outside [0,1] . Since u represents a gene frequency, it is clear in any case that $0 \leq u \leq 1$ is the region of interest. One case of particular interest assumes the local selective advantage s(x) is somewhere positive but is on average negative: $\int_\Omega s(x)dx < 0$. In that case it may be proved [6] that there exists $\lambda_0 > 0$ such that E is simple when $\lambda \neq \lambda_0$,

$$E = \{0,1\} \quad \text{if} \quad 0 < \lambda < \lambda_0$$

$$E = \{0,1,\varphi_\lambda\} \quad \text{if} \quad \lambda > \lambda_0 \quad \text{with} \quad 0 < \varphi_\lambda(x) < 1.$$

Further, the solution $u \equiv 1$ is unstable for all $\lambda > 0$ while $u \equiv 0$ is stable for $0 < \lambda < \lambda_0$ but unstable for $\lambda > \lambda_0$ [5], hence the nontrivial equilibrium φ_λ is stable for $\lambda > \lambda_0$. In fact, we may prove a result somewhat stronger than the Corollary (see [6]): denoting "trivial" the solutions $u \equiv 0$ and $u \equiv 1$, we find every nontrivial solution tends to zero if $0 < \lambda \le \lambda_0$, or to φ_λ if $\lambda > \lambda_0$, as $t \to +\infty$.

REFERENCES.

1. D. Henry, Some gradient dynamical systems defined by parabolic equations, to appear.
2. J. F. G. Auchmuty, Lyapunov methods and equations of parabolic type, Nonlinear Problems, Springer-Verlag, Lect. Notes 322 (1973).
3. D. Henry, Invariant manifolds for maps in Banach spaces, to appear.
4. N. Chafee and E. Infante, A bifurcation problem for a nonlinear P.D.E., Applicable Anal., 4 (1974), 17-37.
5. W. H. Fleming, A selection-migration model in population genetics, J. Math. Biol., 2 (1976), 219-234.
6. D. Henry, A selection-migration model due to W. H. Fleming, to appear.

Daniel Henry
Brown University
Providence, Rhode Island 02912

N J KOPELL

Waves, shocks and target patterns in an oscillating chemical reagent

ABSTRACT

The Belousov reagent is a chemical reagent which oscillates, turning alternately red and blue several times per minute when the appropriate indicator is added and the solution is stirred. When placed in a thin film and left alone, the reagent forms patterns involving regions ("target patterns") of concentric rings which travel outward.

The patterns are complicated and not reproducible in detail. However, certain features of the patterns are repeated with each trial. We isolate out these features and give mathematical descriptions of them in an idealized form. We then look for solutions to reaction-diffusion equations which have these idealized forms. Another part of the program is to find the conditions on the chemical kinetic equations in order that such solutions be possible.

The paper describes some of these "canonical problems." They include periodic plane waves, "slowly-varying waves," shocks separating regions of concentric rings, and idealizations of target patterns. It is shown that the hypothesis that the chemical kinetic equations have a stable periodic solution is sufficient to insure the existence of the plane waves, but not necessarily that of the others (further conditions, one of them on the family of plane waves, are needed). The relationship between mathematics and experimental observations is also discussed.

1. INTRODUCTION

The oscillatory mixture of the title was discovered by Belousov [1], but is better known as the Zhabotinskii [2] reagent. Its ingredients are relatively common: bromate ions, malonic acid, sulfuric acid, and a small amount of cerous ions. However, the result of mixing these chemicals is unusual; the reagent undergoes temporal oscillations with a period of the order of half a minute. With the help of the oxidation-reduction indicator ferroin, the oscillations can be seen as changes in color between bright blue and reddish purple or orange. The mechanism of the oscillation has been extensively investigated, especially by Field, Noyes and Körös [3]. (See the review monograph of Tyson [4] for further references.)

We shall not be concerned here with the details of the mechanism, but rather with the ability of this reagent to spontaneously form patterns. When the fluid is mixed up and placed in a thin (about 2 mm) layer and covered (to prevent convection), certain points in the layer appear to begin to oscillate faster, producing bright blue spots on the darker red background. The blue areas enlarge in size and then turn red at their centers; the red regions in turn enlarge and become blue at the centers. This repeats, and after five or ten minutes the entire surface is covered with sets of outwardly moving concentric rings or "target patterns." A photograph of this configuration may be found in [5] or [6]. (The actual phenomenology is somewhat more complicated than the above. The pattern is not quite uniform throughout the layer, so it is not quite two-dimensional. If there is any fluid motion, the rings can be sheared so that spirals, rather than concentric rings, will form [7]. The three-dimensional version of these spirals are "scrolls" [8]. Even without fluid motion, any gradient in temperature or concentrations of the non-oscillating components alters the observed patterns

[9]. However, all these effects can be minimized, and in this paper we blithely ignore most of them.)

The overall pattern is never completely the same when the experiment is repeated. Nevertheless, there are some features of the patterns which are repeatable. For example, each set of concentric rings has a constant spacing between rings, a constant frequency of oscillation at each point in the target pattern, and (therefore) a constant propagation speed of each ring; all three of these quantities vary from target pattern to target pattern. There is a sharp boundary between any pair of adjacent target patterns across which the oscillation frequency and the spacing of rings changes abruptly. This boundary moves in a predictable way from regions of higher frequency to regions of lower frequency, and the lower frequency regions may be swallowed up completely. (Many of these features were described by Zaikin and Zhabotinskii [2].)

It is generally agreed that only the chemical reaction and diffusion contribute to the formation of the above patterns. It is not well understood what exactly are the properties of the chemical kinetic equations or the diffusion rates which make the patterns possible. The program of L. N. Howard and myself for understanding the patterns has been to examine a number of "canonical problems," each of which is an idealization of a portion of the total pattern. (This is a reasonable thing to do since any given portion evolves quite independently of other portions more than a few millimeters away.) We look for hypotheses on the kinetic equations and diffusion coefficients which are sufficient to solve each of these canonical problems, and which are general enough to include the actual Belousov reagent.

In the rest of this paper, we will discuss some of these problems and the corresponding solutions to the reaction-diffusion equations. These equations have the form

$$c_t = F(c) + K\nabla^2 c \qquad (1)$$

where $c = (c_1,\ldots,c_n)$ is a vector of chemical concentrations, $c_t = F(c)$ are the equations of the chemical kinetics, K is a positive definite matrix of diffusivities, and $\nabla^2 c$ is the spatial Laplacian which models diffusion. The spatial coordinate is called x, and it may have any number of variables; in this paper, x will be one or two dimensional.

2. PLANE WAVES

The simplest idealization of the patterns is a periodic plane wave. That is, if one looks along some radial line of a target pattern with many rings, and one ignores both the boundary of the target pattern and its center, one sees the blue fronts arranged periodically and progressing with fixed speed. Mathematically, this is expressed by saying that, in this region the chemical concentrations are represented by

$$c(x,t) = C(\sigma t - \underline{\alpha} \cdot x) \qquad (2)$$

where C is 2π-periodic in its argument; σ is the frequency, and $\underline{\alpha}$ is the wave vector.

The first question, then, is: under what conditions on the kinetic equations and the diffusion matrix does (1) have solutions of the form (2) ? Even this question has not been answered completely: there are no precisely formulated necessary conditions for the existence of such waves. However, the following sufficient condition is appropriate to the Belousov reagent:

Theorem [5]: Suppose $c_t = F(c)$ has a stable limit cycle solution. Then (1) has a one-parameter family of plane wave solutions which can be parameterized by $\alpha^2 \equiv |\underline{\alpha}|^2$, for small α^2.

Several proofs of this theorem are in [5]. In this paper we present another which is short and conceptual. The main proof in [5] is still more useful for computing the solutions.

<u>Proof</u>: Instead of looking for a 2π-periodic solution of the form (2), we ask for a solution of the form

$$c(x,t) = C(t - \frac{1}{\sigma}\underline{\alpha} \cdot x) \qquad (3)$$

where C is periodic of any period. (Once such a solution is found, it can be rescaled to produce one of the form (2).) Such a C must satisfy an ordinary differential equation:

$$C' = F(C) + \beta K C'' \qquad (4)$$

where $\beta = \alpha^2/\sigma^2$ and $'$ means differentiation with respect to $\zeta \equiv t - (1/\sigma)\underline{\alpha} \cdot x$.

The following argument shows that for each β small enough, there is a periodic solution to (4). We first rewrite (4) as a system:

$$\begin{aligned} C' &= F(C) + U \\ U' &= \beta^{-1}K^{-1}U - M(C)[F(C) + U] \end{aligned} \qquad (5)$$

Here $M(C)$ is a matrix which is the linearization of $F(C)$. ($M_{ij} = \partial F_i/\partial x_j$.) The main idea of the proof is to use the singular perturbation nature of the problem to look for a solution which is close to a solution of $C' = F(C)$, that is, one for which U is uniformly small. We shall do this by showing that (5) has an "invariant manifold" close to $\{U = 0\}$ on which the trajectories of (5) are close to those of $C' = F(C)$.

We rescale (5) by choosing a new independent variable $\overline{\zeta} = -(1/\beta)\zeta$. This turns (5) into:

$$dC/d\bar{\zeta} = -\beta[F(C) + U]$$

$$dU/d\bar{\zeta} = -K^{-1}U + \beta M(C)[F(C) + U] \qquad (6)$$

Now when $\beta = 0$ the system (6) is extremely simple. The manifold $\{U = 0\}$ is an invariant manifold; that is, any trajectory of (6) beginning on the manifold will stay on it for all time. Furthermore, $\{U = 0\}$ is asymptotically stable: trajectories off $\{U = 0\}$ approach it exponentially. Under such circumstances, when the system is perturbed (say by letting β be positive and small) there is a nearby invariant manifold for the perturbed system [10]; it can be expressed as the graph of a function $U = g(C)$ with g small.

(Actually, the theorem in [10] does not strictly hold for $\{U = 0\}$, which is unbounded; the theorem is for finite manifolds with boundary on which the vector field is "overflowing invariant," i.e., the backwards orbit through any point in the manifold is contained in the manifold and the vector field points strictly outward on the boundary. To arrange such a situation in our case, we consider a ball B_1 in C-space (R^n) large enough to contain the limit cycle of $C_t = F(C)$ in its interior. Then we take a larger ball B_2 and change the first equation of (6) by a function of C, only on the annulus between B_1 and B_2, so that the flow points outward on the boundary of B_2 when $\beta = 0$. If this new system is perturbed by letting β be non-zero, the portion of the new invariant manifold lying over B_1 is an invariant manifold for (6). It can be made arbitrarily smooth by making the change on the annulus small enough.)

We now return to (5), armed with our invariant manifold which is also invariant under (5). Inserting $U = g(C)$ into the first equation, we get

$$C' = F(C) + g(C) \qquad (7)$$

on the invariant manifold, with g small; indeed, $g(C)$ can be made arbitrarily small (in the C^1 sense) by letting β be small. Since $C' = F(C)$ has a stable limit cycle, any perturbation of it (such as (7)) also has a periodic solution. This is the required periodic solution to (4), and we have obtained one for each β sufficiently small; it can be shown that this periodic solution is unique.

Remark 1. The above theorem does not require anything of the matrix K but that it be positive definite; however, unless K is close to being scalar, the resulting solutions may be unstable as solutions to (1) [5].

Remark 2. Each of the solutions obtained above may be rewritten in the form (2), using the period of the solution. The σ and the $\underline{\alpha}$ in the expression in (2) are not independent functions of β ; since there is only a one-parameter family of solutions, the σ and $\underline{\alpha}$ are related to one another. That relationship, which we shall write as

$$\sigma = H(\alpha^2) \tag{8}$$

is called the "dispersion relation." The function H depends on F and K. For future reference, we remark here that experiments indicate that the frequency goes up with the absolute value of the wave vector. (In particular, the homogeneous oscillation, which may be thought of as a wave with zero wave number, has a frequency lower than that of any of the target patterns.) Hence the hypothesis "$H' > 0$" is compatible with observation, and "$H' < 0$" is not. As we shall see, "$H' > 0$" also turns out to be a necessary condition to produce solutions which behave like "canonical" parts of the pattern

3. SLOWLY VARYING WAVES

There are no actual plane waves seen in the chemical fluid; the fronts in the target patterns are round. There are other portions of the pattern, especially near the boundaries of the disk or where there has been a disturbance of the fluid, which are neither round nor planar. But the rings, and many of these other cases are locally like plane waves. They can be described by slowly varying waves, which are mathematical creations that model situations in which the space-time behavior of the concentrations can be considered those of a plane wave whose frequency, wave vector and wave form all modulate slowly in space and time. Slowly varing waves are of general interest in the solution of (1) since they give a partial solution to the initial value problem; they describe the evolution of solutions with initial conditions locally close to those of plane waves. In the context of our program to understand the repeatable features of the Zaikin-Zhabotinskii patterns, we are mainly interested in the round target patterns, for which it is the direction of the wave that is varying slowly. (Of course, one must avoid the center of each pattern if this is to be true.)

As we will see, the existence of slowly varying wave solutions to (1) is a direct consequence of the existence of the plane waves. Indeed, it is not even necessary that the chemical kinetic equations have a stable limit cycle; any hypothesis that will produce the plane waves will also yield the slowly varying ones. However, the behavior of the solutions does depend on further structure in the kinetic equations, in particular on the sign of H'. It turns out that the slowly varying waves do not behave like the target patterns unless one assumes $H' > 0$.

The definition of a slowly varying wave is quite technical. Such a "wave" is actually a family of functions of space and time, parameterized by a small parameter ε which, roughly speaking, measures the deviation of the function from being a plane wave. The formal definition is as follows: We first introduce "slow" space and time variables $X = \varepsilon x$, $T = \varepsilon t$.

Definition. A slowly varying wave of order j is a function $c(x,t,\varepsilon)$ which can be expressed in the form:

$$c(x,t,\varepsilon) = Y(\theta,X,T,\varepsilon) \quad \text{with} \quad \theta = \theta(x,t,\varepsilon) \tag{9}$$

where

1. For fixed X,T and ε, Y is a periodic function of θ (the "phase") with least period 2π.

2. $Y(\theta,X,T,\varepsilon)$ has an asymptotic expansion of order j in ε. That is, $Y = Y^0(\theta,X,T) + \varepsilon Y^1 + \ldots + \varepsilon^j Y^j + o(\varepsilon^j)$.

3. If $\Theta(X,T,\varepsilon)$ is defined by $\varepsilon\theta(x,t,\varepsilon) \equiv \Theta(X,T,\varepsilon)$ then Θ has an asymptotic expansion of order $j + 1$ in ε.

The second condition says that the fast variation of Y with x and t is only through the dependence of the variable θ on x and t, so Y is mainly a function of the single variable θ. The third condition says that the "local wave vector" $-\nabla_x\theta \equiv -\nabla_X\Theta$ and the "local frequency" $\theta_t \equiv \Theta_T$ change only on the scales of X and T, so that they are essentially constant over many wave-lengths and many periods; hence the idea of "local wave vector" and "local frequency" makes sense. (See [11] for further discussion of the definition.)

Provided that the periodic functions $Y(\cdot,X,T,\varepsilon)$ are suitably normalized by specifying the zero of phase along the periodic orbit, it can be shown [11] that any function $c(x,t,\varepsilon)$ has at most one representation as a slowly varying wave (of some fixed order j). Furthermore, if $c(x,t,\varepsilon)$ is to be a slowly varying wave and also a solution to (1), then the terms in the asymptotic expansions have to satisfy certain equations. For example, the lowest order term Θ^0 in $\Theta = \Theta^0 + \varepsilon\Theta^1 + \ldots$ must satisfy the equation

$$\Theta^0_T = H(|\nabla_X \Theta^0|^2) \tag{10}$$

where $\sigma = H(\alpha^2)$ is the dispersion function of the plane waves. This says that the waves evolve so that, to lowest order, the local frequency and local wave vector at any X,T are those of some plane wave.

Equation (10) is a first order partial differential equation which can be solved by the method of characteristics. It can be thought of as giving a kinematic description of the propagation of the waves. That is, given $\Theta^0(X,0)$, and hence the local wave vectors at $T = 0$, the equation gives the local frequency; together these determine the direction and speed of propagation of the lines of constant phase, and of the local frequencies and wave vectors at later times. It is a consequence of the structure of equation (1) that the initial variation in phase in fact determines the entire solution, and to all orders: Given $\Theta(X,0,\varepsilon)$, there are unique $\Theta^i(X,T)$ and $Y^i(\theta,X,T)$ which solve (1) in the form (8). $\Theta^0(X,0)$ determines not only $\Theta^0(X,T)$, but also the lowest order part of the wave form, $Y^0(\theta,X,T)$; this turns out, for any given X and T, to be that of the plane wave with frequency Θ^0_T and wave vector $-\nabla_X\Theta^0$. It is in this sense that one can find out the major features of the solution by knowing $\Theta^0(X,0)$ and solving (10).

Equation (10) is easily solved. (See [11].) Its characteristics are straight lines along which Θ^0_T and $-\nabla_X\Theta^0$ are constant. The vector "slope" of the characteristics, $dX/dt = -2\nabla_X\Theta^0 H'(|\nabla_X\Theta^0|^2)$, is the so-called "group velocity" which gives the direction and speed at which changes in

Θ_T^0 and $-\nabla_X\Theta^0$ propagate. When $H' > 0$, this is in the direction of the local wave vector; if $H' < 0$, changes in the local wave vector and frequency travel in the opposite direction from the wave itself.

The round target patterns provide the first instance in which the (experimentally motivated) hypothesis $H' > 0$ is necessary in order that the mathematical solutions mimic the observed behavior in the chemical fluid. The observed waves propagate outward from the center of a pattern. If the center has a fixed frequency, then this frequency is carried outward along the (projection of the) characteristics, and the frequency is constant within the whole target pattern. The slowly varying wave description, which is invalid near the center, does not say how the frequency is maintained there; it does say how the center can appear to act as a pacemaker, setting the frequency far away. This will happen only if $H' > 0$.

4. SHOCKS AND SHOCK STRUCTURES

The slowly varying wave model breaks down near some portions of the overall pattern: it is not valid near the center of each target pattern nor at the boundary of any pattern. However, the kinematic description presented above can be extended, at least to the regions around the boundary of each target pattern, provided that the solutions are not required to be smooth. The basic idea of the kinematics is that the wave motion can be described by giving the evolution of a phase function. This is still true near the boundaries separating any pair of target patterns, or a target pattern from a homogeneously oscillating region. The local wave vector and the local frequency, which are the derivatives of the phase function, do change discontinuously across such a boundary. However, the phase itself continues to make sense, and is continuous across the boundary.

Mathematically, the boundary may be treated as a "shock," a discontinuity in (the derivatives of) a solution to (10) across a surface of codimension one in X,T space. As in the case of some simplified equations for gas dynamics, if one allows discontinuous solutions, one must give conditions governing the evolution of the discontinuity; otherwise the solution is not uniquely specified. In gas dynamics and other similar situations, these conditions ("Rankine-Hugoniot conditions") come from the physical principles underlying the partial differential equation, such as conservation of mass, momentum and energy. In our case, the evolution of the shock is determined by requiring that the phase function continue to be defined and continuous across the shock. This turns out to force the shock to move normal to itself at the speed $\Delta\sigma/\Delta\alpha$, where $\Delta\sigma$ is the difference of the frequencies on the two sides and $\Delta\alpha$ is the difference of the wave vectors in the direction normal to the shock. This is exactly the speed that makes equal the Doppler-shifted frequencies on the two sides.

A real shock in gas dynamics must satisfy more than the Rankine-Hugoniot conditions; there is also an entropy condition to be fulfilled. Similarly, in addition to the "phase continuity rule," we need an analogue to the entropy condition. This is the requirement that characteristics of (10) on the two sides of the discontinuity front must converge on the front as T increases. Such a rule is necessary in order that the region of discontinuity remain "thin" (i.e. codimension one); otherwise, as time increases, the region over which there is a large change in local wave vector will grow.

We note here another instance in which $H' > 0$ turns up as a requirement for the mathematical description. Consider the situation of two target patterns with rings propagating toward the front separating them. Then the characteristics of the mathematical solution corresponding to this situation will converge on the discontinuity surface if and only if $H' > 0$. That is, the physical shocks (i.e. waves of opposing direction converging on one another) are mathematical shocks (i.e. satisfy an entropy condition) only if $H' > 0$.

The simplified equation (10), plus shock conditions, predicts the evolution of the phase and also of the shocks. However, these rules take the existence of shocks as given; the underlying hypothesis is that the full equation (1) has solutions in which there are rapid transitions ("shock structure") in the local wave vector and frequency over a localized transition region which, in the appropriate limit, becomes the shock surface. This hypothesis requires mathematical verification.

The simplest example of a shock is that of a pair of plain waves converging upon one another, with colinear wave vectors. (This is an idealization of the portion of the pattern around a straight line joining the centers of two adjacent target patterns, which are conceived of as infinitely large, so the centers of the two patterns are at infinity.) Even for this idealization, the problem has been solved rigorously only for a class of model equations for which the shock structure problem can be formulated as a question about ordinary differential equations. An example of such model equations for the chemical kinetics is

$$c_t = \begin{pmatrix} \lambda & -\omega \\ \omega & \lambda \end{pmatrix} c \tag{11}$$

where $c = (c_1, c_2)$, $\lambda(c) = 1 - |c|^2$ and $\omega(c) = 2 - |c|^2$. (These equations have a stable limit cycle at $|c| = 1$. The dispersion function H can be computed: it is $H(\alpha^2) = \omega \circ \lambda^{-1}(\alpha^2) = 1 + \alpha^2$, so $H' = 1$.) For simplicity, we describe only the problem involving equal and opposite waves, in which the shock front is stationary. (See [11] for other cases.) We also assume that x is one-dimensional (which turns out to be no loss of generality for this question).

Let r,θ be defined by $c_1 = r\cos\theta$, $c_2 = r\sin\theta$ (so r,θ are polar coordinates in concentration space.) If we take $K = I$, equation (1) in r,θ coordinates is

$$\begin{aligned} r_t &= r\lambda(r) + r_{xx} - r|\theta_x|^2 \\ \theta_t &= \omega(r) + \frac{1}{r^2}(r^2\theta_x)_x \end{aligned} \tag{12}$$

We want a solution to (12) which asymptotically approaches plane waves as $x \to \pm\infty$. The plane waves of (12) can easily be written down. They are:

$$\begin{aligned} r &= r_0 \\ \theta &= \sigma t - \alpha x + \theta_0 \end{aligned} \tag{13}$$

where $\sigma = \omega(r_0)$ and $\alpha^2 = \lambda(r_0)$. Here α means the component of the one-dimensional wave vector $\underline{\alpha}$, so α can be positive or negative. (Since λ can be inverted, this family can be parameterized by α^2 as stated earlier for general kinetic equations.)

Now we also wish our shock structure solution to have some "permanence" in time; for example, in the equations of gas dynamics, the shock structure is time independent in the appropriate moving coordinate system. Time independence cannot be achieved in our problem because the asymptotic states are oscillatory. However, we can ask that the amplitude, r, of the solution (but not the phase, θ) be independent of time. The structure of (12) then forces θ to have a simple form: θ_t is a constant (independent of x and t) which we may call σ, so r and θ may be written as

$$r = r(x)$$
$$\theta = \sigma t - \int^{x} a(x')dx' \tag{14}$$

If we insert (14) into (12), we get ordinary differential equations for $r(x)$ and $a(x)$ which can be written as a system:

$$\begin{aligned} r_x &= ru \\ u_x &= -\lambda(r) + a^2 - u^2 \\ a_x &= \omega(r) - \sigma - 2au \end{aligned} \tag{15}$$

Equation (15) (which is actually a one-parameter family of equations depending on σ) has a pair of critical points: $(r_0, 0, \pm\,\alpha)$, where $\omega(r_0) = \sigma$ and $\lambda(r_0) = \alpha^2$. These critical points represent the plane waves which are the asymptotic states to be connected. (The plane waves have constant r, constant frequency $\theta_t = \sigma$ and constant wave number $-\theta_x = a$.) Since we wish the plane waves to be converging, the solution we require has a positive wave number for $x \to -\infty$ and a negative wave number for $x \to +\infty$. At least for σ close to and greater than the frequency of the limit cycle oscillation of (11), there is a unique trajectory of (15) which tends to $(r_0, 0, \mp\,\alpha)$, $\alpha > 0$, as $x \to \pm\infty$. This solution can be found by bifurcation methods [11, 12] since, as $\sigma \to 1$, the critical points of (15) coalesce. The proof still holds if the functions λ and ω are changed, but it is crucial that λ' and ω' remain negative (so H' is positive).

For more general kinetic equations, as long as there is a one parameter family of plane wave solutions and $H' > 0$, the shock problem can be turned into a problem which formally again requires finding a trajectory which joins two critical points [11]. As before, the hypotheses of the relevant bifurcation theorem in [12] are satisfied, this time with one exception: the space is not finite dimensional. In the general case, the critical points are in a function space and the "differential equation" is not defined on the whole function space. Hence there are some tricky technicalities in infinite dimensional analysis to be overcome in order to turn the formal treatment into a rigorous proof.

5. TARGET PATTERN SOLUTIONS

The final problem to be considered deals with another region in which the slowly varying wave model breaks down, namely the center of each target pattern. In this problem, an additional hypothesis appears to be needed in order to get a solution.

We shall look for solutions to (1) representing a single target pattern (including the center) having infinite extent. This is an idealization of a target pattern with many rings; the furthest ring is pushed to infinity. Such solutions are radially symmetric and asymptotic to outgoing waves for large radius. Like the shock structure question, this problem has so far been solved only for kinetic equations having the special form (11). Indeed, there is not even a formal analogue yet available for more general kinetic equations. But even for the special equations (11), a sharp contrast emerges between the shock structure problem and the present question which will be called "the target pattern problem."

We consider first a one-dimensional version of the target pattern problem. (Think of the region around the diameter of a two-dimensional target pattern, or of carrying out the experiment in a long thin tube.) The problem is then to find a solution which asymptotically approaches <u>outgoing</u> plane waves of equal and opposite wave numbers; it is identical to the shock problem with the boundary conditions reversed. Hence, we shall again look for solutions

$r(x,t)$, $\theta(x,t)$ for which the amplitude is time independent, and we are again led to (15). This time our trajectory must approach $(r_0,0,\pm\alpha)$ as $x \to \pm\infty$.

The first indication of the difference between the problems comes when one computes the dimensions of the space of trajectories leaving and entering each of the critical points. When $H' > 0$, the dimension of the stable manifold of $(r_0,0,-\alpha)$ is two, and that of its unstable manifold is one; these dimensions are reversed for $(r_0,0,+\alpha)$. Thus if there is to be a trajectory from $(r_0,0,-\alpha)$ to $(r_0,0,+\alpha)$, the unique trajectory leaving $(r_0,0,-\alpha)$ must coincide with the unique trajectory entering $(r_0,0,+\alpha)$, a highly unlikely situation in three-dimensional space. Nevertheless, it turns out that there are such solutions, at least for some discrete values of σ ; unlike the shock problem, which has a solution for a continuous range of σ, the target pattern problem is a nonlinear eigenvalue problem with eigenvalue σ (or r_0, where $\omega(r_0) = \sigma$).

The existence of these solutions has been shown rigorously only under a further restriction on the kinetic equations. This restriction is that ω' be "small." That is, if $\omega(r)$ is scaled by $\omega(r) - \omega(r_0) \equiv \varepsilon[\bar{\omega}(r) - \bar{\omega}(r_0)]$ where $\bar{\omega}(r) - \bar{\omega}(r_0) = 0(1)$, then ε should be small. The mathematical motivation for considering this case is that the dimensions of the stable and unstable manifolds of the critical points change if ω' is changed from negative to positive; hence some interesting phenomenon can be expected to happen at $\omega' \equiv 0$, which might be used to study the case of ω' small. This turns out to be true, and one can prove:

<u>Theorem [13]</u>: For ε sufficiently small, there are discrete values of r_0 for which (15) has a solution which goes from $(r_0,0,-\alpha)$ to $(r_0,0,+\alpha)$. The number of these values can be made arbitrarily large by making ε small

enough.

The proof of the theorem actually demonstrates more than is stated above. It shows that there are two different types of solutions to the problem. Half of them are symmetric in the original (c_1, c_2) coordinates, with $r_x = a = u = 0$ at $x = 0$; the other half are antisymmetric. (For the latter solutions, $r = 0$ at $x = 0$, and θ and r_x cease to be defined there.) The second type of solution may be thought of as a one-dimensional version of a spiral, with waves emanating from the center alternately on the two sides rather than synchronously.

The major ideas of the proof are as follows: If one sets $\varepsilon = 0$, then (15) can be interpreted as a central force problem. That is, the equations are:

$$
\begin{aligned}
&r_{xx} - r\theta_x^2 = -r\lambda(r) \\
&r\theta_{xx} + 2\theta_x r_x = 0
\end{aligned}
\tag{16}
$$

The left hand sides represent the radial and tangential components of acceleration of a particle moving in a force field; according to (16), this is a central force field, with the central force related to $\lambda(r)$. This set of equations has a pair of integrals representing conservation of angular momentum and energy. These are

$$j = r^2 a$$

$$e = \frac{1}{2}\left[r^2(a^2 + u^2) + g(r)\right]$$

where $g'(r) = 2r\lambda(r)$ and $g(1) = 0$.

One rewrites (15) in variables suited to the integrals j and e, and scales in the following way: $R \equiv r^2$, $V \equiv R_x$, $J \equiv j/\varepsilon$, $E \equiv e/\varepsilon^2$,

$G(R) \equiv g(r)$, $\Omega(R) = \overline{\omega}(r)$. We also introduce a parameter S to replace the parameter r_0 ; S is defined by $S\varepsilon^2 = \Omega(R_0) - \Omega(1)$. Equations (15) then become:

$$R_x = V$$

$$V_x = 4\varepsilon^2 E - 2(RG'(R) + G)$$

$$J_x = -R[\Omega(1) - \Omega(R) + S\varepsilon^2]$$

where $4\varepsilon^2 E = (\frac{1}{2}V^2 + 2\varepsilon^2 J^2)/R + 2G$. (The scaling for J is chosen so that, as $\varepsilon \to 0$, the critical points stay bounded away from each other and infinity. The variable e is scaled as above since it turns out to be of order ε^2.)

When $\varepsilon = 0$, the first two equations of (17) decouple from the third. (We shall refer to these first two as $(17)_0$.) They have a phase-plane diagram that is shown in Figure 1. In particular, there is a critical point at $R = 1$, $V = 0$. When ε is close to zero, (17) has a pair of critical points $(R_0, 0, \pm J_0)$ whose (R,V) components are close to $(1,0)$; the trajectory we are seeking goes from $(R_0, 0, -J_0)$ to $(R_0, 0, +J_0)$, $J_0 > 0$. In addition, it turns out that, for small ε, the (R,V) components of the vector field (17) are close to those of $(17)_0$ in a neighborhood of the unstable manifold of $(R_0, 0, -J_0)$, and this can be used to follow the unstable manifold.

One now uses the estimates on the unstable manifold to show that $S = S_i(\varepsilon)$ can be chosen so that the unstable manifold goes through $V = 0$, $J = 0$. Then, by the symmetry of the equations, the trajectory must approach $(R_0, 0, J_0)$ as $x \to +\infty$. This unstable manifold may hit $V = 0$, $J = 0$ either near $R = 0$ or near $R = 1$ (see Figure 1); the subscript i represents

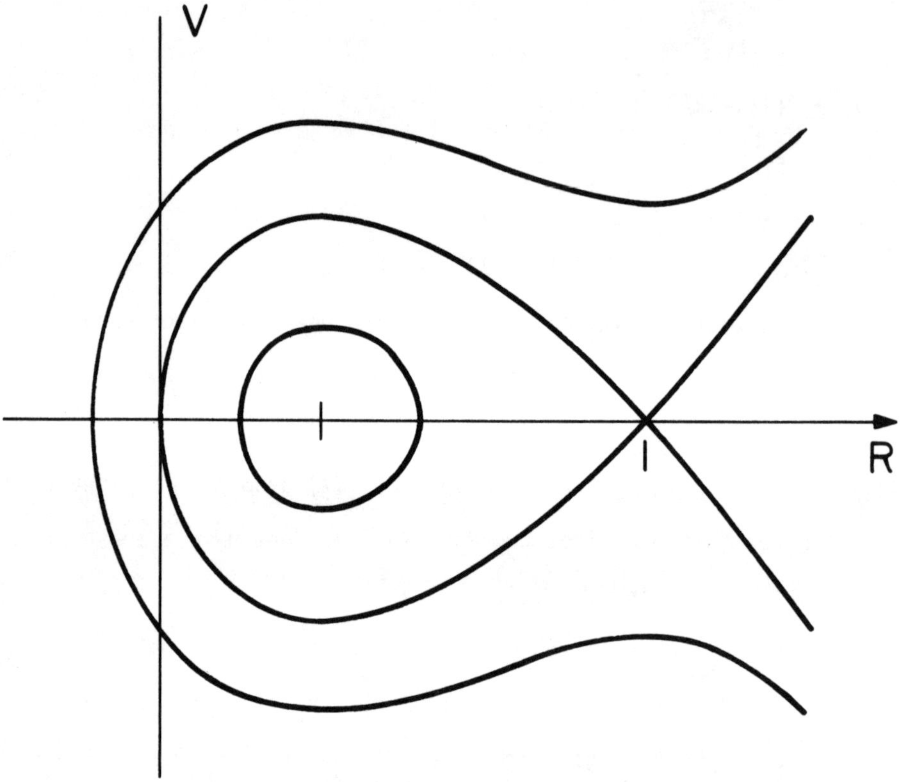

Figure 1

The integral curves of $R_x = V$, $V_x = -2(RG'(R) + G)$.

They are the level curves of the integral $\frac{1}{2}V^2 + 2RG$.

the number of half times around (near) the homoclinic orbit of $(17)_0$ that the unstable manifold travels before hitting $V = 0$, $J = 0$. It can be shown that, for all odd i, the trajectory actually hits $R = 0$, $V = 0$, $J = 0$; these are the "spiral-like" solutions. The even values of i give the symmetric solutions. The larger the value of i, the smaller ε must be taken to insure the existence of the solution. The proof does not reveal whether for fixed ε, the number of solutions is finite.

A similar analysis can be tried for the two dimensional version of the center problem. In that case, we must replace the r_{xx}, θ_{xx} and θ_x in (12) by $\nabla^2 r$, $\nabla^2\theta$ and $\nabla\theta$. If we now denote by x the radial distance from the center and again look for solutions of the form (14), we get the non-autonomous system

$$
\begin{aligned}
r_x &= ru \\
u_x &= -\lambda(r) + a^2 - u^2 - u/x \\
a_x &= \omega(r) - \sigma - 2au - a/x \, .
\end{aligned}
\tag{18}
$$

This time the analysis is somewhat more delicate, and the work is not yet finished. It can still be shown that there are at most discretely many values of σ for which there is a trajectory of (18) satisfying $(r,u,a) \to (r_0,0,\alpha)$ as $x \to \infty$ and $u = a = 0$ at $x = 0$; formal calculations using R,V,J coordinates indicate that there are such solutions, at least for some functions $\lambda(r)$ and $\omega(r)$.

Similarly, one can look for spiral solutions.

$$
\begin{aligned}
r &= r(x) \\
\theta &= \sigma t + m\varphi - \int^x a(x')dx'
\end{aligned}
\tag{19}
$$

where φ is the angular variable in (x_1,x_2) space (so $x_1 = x \cos \varphi$, $x_2 = x \sin \varphi$). When (19) is inserted in (the multidimensional version of) (12), the φ drops out, and again one gets a non-autonomous ordinary differential equation in x, this time dependent on m. The analysis of this equation is similar to that of (18), and is also not yet complete.

All the analysis of this section depends crucially on the hypothesis that ε is small. There is some (but not yet definitive evidence) that there might be no solutions to the target pattern problem if ε is too large; in any case it can be shown that, for ε too large, whatever solution exists must be unstable [13]. Thus, the analysis of the target pattern problem shows that the hypothesis sufficient to produce solutions to the other canonical problems are not quite enough to yield a solution to the target pattern (or spiral) problem. It is still unclear how to interpret the "small ε" condition for more general kinetic equations having a stable limit cycle, so it is not yet obvious even how to formulate a generalization of this section.

6. FURTHER COMMENTS AND CONCLUSIONS

The analysis of the canonical problems shows that it is not necessary to have detailed knowledge of the kinetic equations in order to understand why the patterns form. We have seen that many of the major features of the overall patterns are consequences of a few simple hypotheses on the kinetic equations. (It is possible that by considering other features of the patterns, one might uncover more necessary conditions.)

If it is true, under some appropriate hypothesis, that general kinetic equations do behave qualitatively like (11) (providing they have a stable limit cycle and the resulting dispersion relation H satisfies $H' > 0$), and if the two dimensional cases are indeed qualitatively like the one dimensional case, then the theory also predicts that the frequencies of the target patterns should be discrete. At first glance, this seems to be experimentally contradicted; when the experiment is carried out as described above, there is a range of wave numbers, with no evidence of discreteness. However, that is misleading because of the possible existence of foreign particles in the fluid. Winfree [14] reports that if the liquid is first carefully filtered, the number of resulting target patterns can be cut down or even eliminated. (This claim is not without controversy [15].) Winfree reports further that if spirals are caused to start in this carefully filtered fluid by poking the fluid with a hot needle and shearing the fluid, then all the resulting spirals have the same frequency [14].

These observations can be interpreted in a simple way within the above theory. The homogeneous fluid is capable of supporting a one-parameter family of plane waves, with a range of possible frequencies. However (at least if a generalization of Section 5 is true) only a discrete number (and possibly only one) of these can be the asymptotic form of a target pattern or spiral solution having the boundary condition which corresponds to the absence of a foreign particle at the center. Suppose now that there is a foreign particle at $x = 0$ which acts to change the kinetics locally around that point. For a target pattern, this can be expressed mathematically by changing the boundary condition at $x = 0$ from $c_x = 0$ to $c_x = h(x)$ for some function $h(x)$. It can be shown [13], at least for h small and $c_t = F(c)$ of the form (11), that the solutions to the target pattern problem are still discrete, but the frequencies involved are changed. Thus, if there are many foreign particles in the fluid, each contributing a somewhat different boundary condition, the frequencies expressed need no longer appear to be discrete. If, however, the particles are filtered out, it is not surprising that there may be only a single frequency expressed.

The hypotheses of the theory, as general as they are, are still not necessary conditions. More specifically, it seems probable that one can generalize the hypothesis that $c_t = F(c)$ has a stable limit cycle. Work on variations of the Nagumo equations [16-18] suggests that if $c_t = F(c)$ has a solution which is "close" to being a limit cycle (i.e. one which starts close to a critical point and goes far away before returning to the critical point), then (1) may still have a one-parameter family of plane waves; but no such general theorem has been proved. Experimental work also suggests that there should be such a generalization: a small change in the initial concentrations of the reagent turns it into a fluid which does not oscillate, but which is still capable of supporting traveling waves [7].

Another important problem which remains open is that of finding a mathematical description of how the target patterns begin to form. (The above theory describes the evolution of the configurations after a relatively long time.) Ideally, such a description should be independent of specific models and based on testable properties of the chemical oscillators such as the Belousov reagent.

REFERENCES

1. B. Belousov, Sb. ref. radiats. med. Medzig, Moscow (1959).
2. A. N. Zaikin and A. M. Zhabotinskii, Concentration Wave Propagation in Two-dimensional Liquid-phase Self-oscillating System, Nature 225, 535 (1970).
3. R. M. Noyes, R. J. Field and E. Körös, Oscillations in Chemical Systems. II, J. Am. Chem. Soc. 94, 8649 (1972).

4. J. Tyson, The Belousov-Zhabotinskii Reaction, Lecture Notes in Mathematics, ed. S. Levin, 10, Springer-Verlag, New York (1976).

5. N. Kopell and L. N. Howard, Plane Wave Solutions to Reaction-Diffusion Equations, Studies in Appl. Math., 52, 291 (1973).

6. L. N. Howard, Bifurcation in Reaction-Diffusion Problems, Advances in Mathematics, 16, 245, (1975).

7. A. T. Winfree, Spiral Waves of Chemical Activity, Science, 175, 634 (1972).

8. A. T. Winfree, Scroll-Shaped Waves of Chemical Activity in Three Dimensions, Science, 181, 937 (1973).

9. N. Kopell and L. N. Howard, Horizontal Bands in the Belousov Reaction, Science, 180, 1171 (1973).

10. N. Fenichel, Persistence and Smoothness of Invariant Manifolds for Flows, Indiana U. Math. J., 21, 193 (1971).

11. L. N. Howard and N. Kopell, Slowly Varying Waves and Shock Structures in Reaction-Diffusion Equations, Studies in Applied Math., to appear (1977).

12. N. Kopell and L. N. Howard, Bifurcations and Trajectories Joining Critical Points, Advances in Mathematics 18, 306, 1975.

13. L. N. Howard and N. Kopell, Target Pattern Solutions to Reaction-Diffusion Equations: a Perturbed Central Force Problem, in preparation.

14. A. T. Winfree, Wavelike Activity in Biological and Chemical Media, Lecture Notes in Biomathematics, ed. P. van der Driessche, Springer-Verlag, (1974).

15. A. M. Zhabotinskii and A. N. Zaiken, Autowave Processes in a Distributed Chemical System, J. Theor. Biol. 40, 45 (1973).

16. G. Carpenter, A Geometric Approach to Singular Perturbation Problems with Applications to Nerve Impulse Equations, J. Diff. Equa., to appear (1977).

17. J. Rinzel and J. Keller, Traveling Wave Solutions of a Nerve Conduction Equation, Biophysical J., 13, 1313 (1973).

18. A. T. Winfree, Rotating Solutions to Reaction-Diffusion Equations in Simply-Connected Media, SIAM-AMS Proc., 8, 13 (1974).

Nancy J. Kopell
Department of Mathematics
Northeastern University
Boston, Massachusetts 02115

R M MIURA

A nonlinear WKB method and slowly-modulated oscillations in nonlinear diffusion processes

ABSTRACT

In this paper two topics are discussed. The first discussion gives a qualitative and intuitive picture of a nonlinear WKB method which has been designed and used to obtain approximations to slowly-varying solutions of nonlinear partial differential equations. In the second discussion, approximate slowly-modulated oscillatory solutions for a certain class of reaction-diffusion equations are obtained by a formal asymptotic procedure. These solutions describe phenomena in chemical reactors, chemical and biological reactions, and in other media where a stable oscillation at each space point undergoes a slow amplitude change due to diffusion. The leading-order solutions take the form $R(\xi - c\tau)P(t^*)$ where $P(t^*)$ is a local sinusoidal oscillation on a fast time scale t^* and $R(\xi - c\tau)$ is a steady progressing wave which slowly modulates the amplitude on slow space (ξ) and slow time (τ) scales. Depending on the speed of propagation c, which is undetermined, qualitatively different modulations are obtained.

1. INTRODUCTION

Several different methods for obtaining approximate solutions of nonlinear partial differential equations arising in physical problems have been developed recently. These methods take into account at leading order the full nonlinearity of the equations and hence do not have the disadvantages of some nonlinear perturbation methods which begin with linear approximations. Obviously, for nonlinear equations, the loss of a superposition principle prevents the use of fundamental solutions from which other solutions can be constructed.

In the study of nonlinear waves in dispersive and dissipative media, there are a variety of wave phenomena when a small parameter appears in the problem including waves which are nearly periodic. By nearly-periodic we mean that the waves oscillate but with slowly changing frequency and amplitude -- these changes being small over many "periods" of the waves. Examples of equations exhibiting such wave solutions are the Korteweg-deVries equation

$$u_t + uu_x + \delta^2 u_{xxx} = 0 , \quad \delta^2 << 1 , \tag{1.1}$$

in the dispersive case, and nonlinear reaction-diffusion equations of the type studied by Howard and Kopell [1] - [3] or studied later in this paper.

In both these cases, we take advantage of the existence of a small parameter in the model equations to devise solution methods using perturbation theory.

From the computational point of view, these methods are also very useful. Computer generated solutions become difficult to obtain since the oscillations become more closely spaced as the small parameter decreases and hence accurate computations require more space steps. Because of practical limitations such as stability and accuracy of the numerical scheme, one becomes severely limited in the total computation time.

Generally, to achieve any explicit results for nonlinear problems, one must restrict the class of solutions being sought. In this paper this restriction will be to slowly-modulated oscillatory solutions of certain nonlinear partial differential equations.

First, a qualitative and intuitive description will be given of a nonlinear WKB method which was developed for obtaining approximate slowly-modulated oscillatory solutions of nonlinear dispersive wave problems [4] - [8]. This discussion is intended to aid the uninitiated reader in understanding some of the procedures used in the applications of that method.

In the remainder of the paper we will describe a formal asymptotic procedure for obtaining approximate slowly-modulated solutions for a certain class of coupled reaction-diffusion equations. These types of solutions describe phenomena in chemical reactors, chemical and biological reactions, and in other media where a local stable oscillation undergoes a slow amplitude change due to diffusion.

2. A NONLINEAR WKB METHOD - A DESCRIPTION

In another paper in this volume, see Kopell [9], a description is given for finding modulated periodic waves for reaction-diffusion systems. Here we give a qualitative description of a nonlinear WKB method which serves to give a different point of view of the analysis by Kopell.

To avoid complications, we consider a second-order ordinary differential equation in the general form

$$F(x,u_{xx},u_x,u;\varepsilon) = 0 \ , \ \varepsilon << 1 \ , \tag{2.1}$$

which has solutions $u(x;\varepsilon)$ like those shown in Figure 1. For example, the linear ordinary differential equation

$$\varepsilon^2 u_{xx} + V(x)u = 0 \ , \tag{2.2}$$

where $V(x)$ is slowly varying on a length scale $O(\varepsilon)$ exhibits the desired behavior. Of course, this can be analyzed by the linear WKB method. For the linear WKB method, the essential step is to assume a solution in the form

$$u(x;\varepsilon) \sim U(x,\Theta;\varepsilon) \equiv W(x;\varepsilon)e^{i\Theta} \quad , \quad \Theta \equiv \frac{B(x;\varepsilon)}{\varepsilon} \quad , \tag{2.3}$$

where there are now two new functions $W(x;\varepsilon)$ and $B(x;\varepsilon)$ instead of one function u ; therefore, we have the freedom to specify one condition between them. The functions W and B are slowly varying functions on a length scale $O(\varepsilon)$.

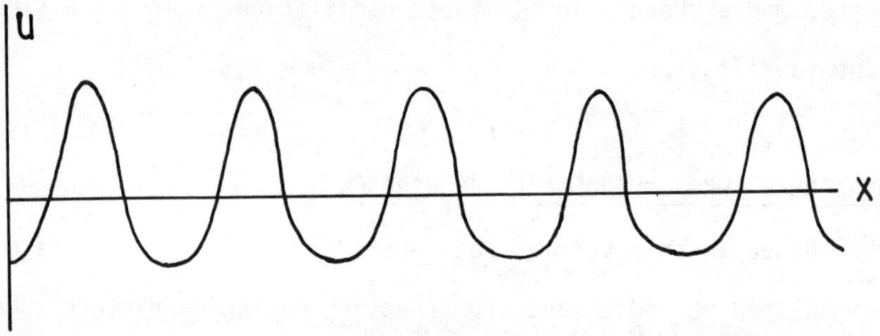

Figure 1

Our objective is to obtain an approximation to $u(x;\varepsilon)$ which reflects both the near periodicity and the slow modulation in x. Note that the basic oscillation need not be close to sinusoidal as in the linear case.

We begin by a formal extension of the problem into an independent variable space of one higher dimension, i.e., let $u(x;\varepsilon) \to U(x,\Theta;\varepsilon)$ where we impose the conditions that U be strictly periodic in the variable Θ and have period 1 in that variable; thus

$$U(x,\Theta;\varepsilon) = U(x,\Theta + 1;\varepsilon) \quad . \tag{2.4}$$

The governing partial differential equation for U is obtained by treating $\Theta = \Theta(x;\varepsilon)$ and rewriting the original equation (2.1) by appropriate use of the chain rule. The quantity Θ_x is defined as a new variable to be determined in solving the extended problem. This dependence of Θ on x and ε generally takes the form

$$\Theta(x;\varepsilon) = \frac{B(x;\varepsilon)}{\varphi(\varepsilon)} \quad ; \quad B(x;\varepsilon) = O(1),\ \varphi(\varepsilon) = o(1)\ . \tag{2.5}$$

The reason for the choice of the singular behavior in ε is to ensure a rapid oscillation as $\varepsilon \to 0$. Thus one arrives at a new equation in the form

$$F(x,\Theta_x,U_{xx},U_{x\Theta},U_{\Theta\Theta},U_x,U_\Theta,U;\varepsilon) = 0\ . \tag{2.6}$$

Although this problem looks more formidable than the original problem, all the variables and derivatives are $O(1)$ in ε and hence a regular perturbation procedure can be followed in computing successive approximations to $U(x,\Theta;\varepsilon)$ and to $\Theta(x;\varepsilon)$. If these can be found we obtain a function of x and ε

$$U(x,\Theta(x;\varepsilon);\varepsilon) \tag{2.7}$$

which is supposed to represent an approximation to the original function $u(x;\varepsilon)$. This procedure has been called the "method of extension" by Sandri [10]. The difficult problem is this last step, namely to show that $U(x,\Theta(x;\varepsilon);\varepsilon)$ is asymptotic to $u(x;\varepsilon)$.

From Figure 2, we can see geometrically why the nonlinear WKB method leads to a simplification in the approximate construction of the nearly-periodic solutions. The surface $U(x,\Theta)$ (we suppress the dependence on ε for this discussion) is drawn in (x,Θ,U)-space. In the (x,Θ)-plane we have the relationship $\Theta = \Theta(x)$ which is projected vertically to intersect the $U(x,\Theta)$ surface. The intersection of these two surfaces represents the desired solution $U(x,\Theta(x)) \sim u(x)$. The near periodicity of the solution

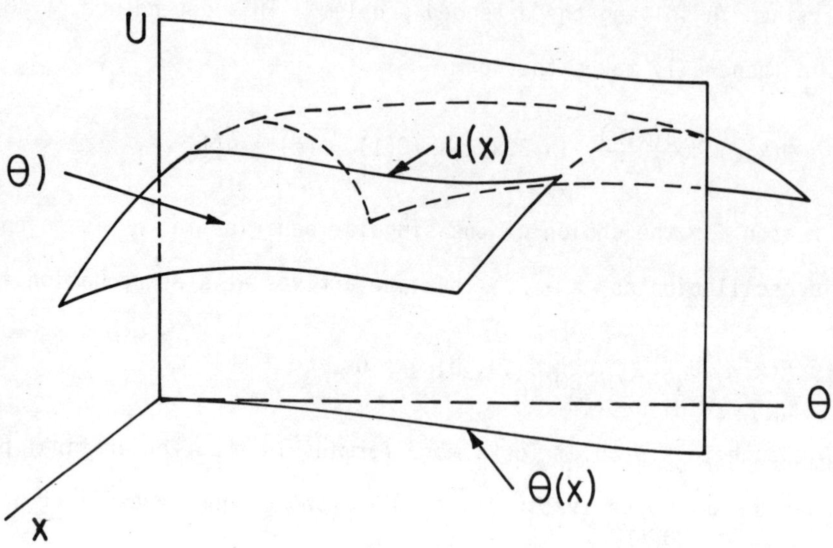

Figure 2

occurs since the curve $\Theta(x)$ lies very close to the Θ-axis for small ε. Thus a small change in x corresponds to a large change in Θ. This means that Θ goes through many periods of U, whereas the amplitude and frequency of these oscillations (which depend on x) vary little since x has hardly changed.

3. SLOWLY-MODULATED OSCILLATIONS IN NONLINEAR DIFFUSION PROCESSES

Under certain circumstances in a chemical reactor or a chemical or biological reaction, the usual stationary state which is oscillatory or periodic at each point in space will undergo a slow change or drift. Analogous phenomena occur in ecological systems. Here the objective is to find approximate slowly-varying oscillatory wave solutions of the model equations which incorporate the effects of both diffusion and nonlinearity.

Although a more general class of equations can be treated, here we only study the system of equations

$$\begin{aligned} u_t &= D_1 u_{xx} + \alpha(\lambda)u - \beta v - \lambda u^3 , \\ v_t &= D_2 v_{xx} + \beta u + \gamma(\lambda)v - \mu\lambda u^2 v \end{aligned} \tag{3.1}$$

A more detailed treatment is contained in Cohen, Hoppensteadt and Miura [11]. This system is used to model first-order tubular chemical reactions when $\alpha = \gamma$. For a different choice of the parameters one gets the Nagumo equation in the regular oscillation case

$$\begin{aligned} U_t &= U_{xx} + \varepsilon^2(U - \frac{1}{3}U^3) - V , \\ V_t &= U \quad . \end{aligned} \tag{3.2}$$

Our objective will be to look for small amplitude solutions which consist of a local sinusoidal oscillation that is slowly modulated by a steady progressing wave.

For the purposes of this paper we consider parameter values near a bifurcation point $\lambda = \lambda_0$. For $\lambda = \lambda_0$ we assume that the solution of interest is a pure linear oscillation about the static zero state, whereas for $\lambda = \lambda_0 + \varepsilon$, $\varepsilon > 0$, the diffusionless model system of equations is linearly unstable.

Assume that the parameters $\alpha(\lambda)$ and $\gamma(\lambda)$ are given by

$$\begin{aligned} \alpha(\lambda) &= \varepsilon^2\alpha_2 + \varepsilon^3\hat{\alpha} , \\ \gamma(\lambda) &= \varepsilon^2\gamma_2 + \varepsilon^3\hat{\gamma} , \end{aligned} \qquad \alpha_2 + \gamma_2 > 0 , \quad 0 < \varepsilon \ll 1 , \tag{3.3}$$

where α_2 and γ_2 are constants independent of ε and $\hat{\alpha}$ and $\hat{\gamma}$ are both

$o(1)$ in ε. We know that the linear equations

$$u_t = -\beta v \ ,$$
$$v_t = \beta u \ , \tag{3.4}$$

lead to sinusoidal solutions, so the equations (3.1) are obtained from these by adding diffusion and nonlinear effects and destabilizing linear terms. Let the desired solutions be of small amplitude given by

$$u = u(\eta,t^*;\varepsilon) = \varepsilon F(\eta,t^*;\varepsilon) \ ,$$
$$v = v(\eta,t^*;\varepsilon) = \varepsilon G(\eta,t^*;\varepsilon) \ , \tag{3.5}$$

where we have defined the new variables

$$\eta \equiv \xi - c\tau = \varepsilon x - c\varepsilon^2 t \ ,$$
$$t^* \equiv (1 - \varepsilon\omega(\varepsilon))^{-1} t \ . \tag{3.6}$$

The variable η corresponds to a steady slow translation variable of a steady progressing wave moving with velocity c in slow space and slow time coordinates, ξ and τ, respectively. The variable t^* is a fast time scale and the scale factor essentially corresponds to a Poincaré correction to the frequency of the fast time scale oscillation.

Substitution of these changes of variables into the original equations yields the governing equations

$$F_{t*} + \beta G = \varepsilon\omega\beta G + \varepsilon^2 \mathcal{F}(F,F_\eta,F_{\eta\eta};\varepsilon) \ ,$$
$$G_{t*} - \beta F = -\varepsilon\omega\beta F + \varepsilon^2 \mathcal{G}(F,G,G_\eta,G_{\eta\eta};\varepsilon) \ , \tag{3.7}$$

where

$$F = (1 - \varepsilon\omega)[D_1 F_{\eta\eta} + cF_\eta + (\alpha_2 + \varepsilon\hat{\alpha})F - (\lambda_0 + \varepsilon)F^3] ,$$
$$G = (1 - \varepsilon\omega)[D_2 G_{\eta\eta} + cG_\eta + (\gamma_2 + \varepsilon\hat{\gamma})G - \mu(\lambda_0 + \varepsilon)F^2 G] . \tag{3.8}$$

Furthermore, we assume that the quantities $\omega(\varepsilon)$, $F(\eta,t^*;\varepsilon)$, and $G(\eta,t^*;\varepsilon)$ can be expanded in ε as

$$\omega(\varepsilon) = \varepsilon\omega_2 + \varepsilon^3\omega_3 + \dots ,$$
$$F = F_0 + \varepsilon^2 F_2 + \varepsilon^3 F_3 + \dots , \tag{3.9}$$
$$G = G_0 + \varepsilon^2 G_2 + \varepsilon^3 G_3 + \dots ,$$

To leading order in ε we obtain

$$F_{0_{t^*}} + \beta G_0 = 0 ,$$
$$G_{0_{t^*}} - \beta F_0 = 0 , \tag{3.10}$$

with general solutions

$$F_0(\eta,t^*) = R(\eta)\cos(\beta t^* + \phi(\eta)) ,$$
$$G_0(\eta,t^*) = R(\eta)\sin(\beta t^* + \phi(\eta)) . \tag{3.11}$$

We now determine the functions $R(\eta)$ and $\phi(\eta)$ from consistency conditions for the solvability of the equations governing F_2 and G_2. The equations for F_2 and G_2 are linear and inhomogeneous, and to avoid secular terms at this order in ε, orthogonality conditions on the inhomogeneous terms yield the following ordinary differential equations for R and ϕ

$$D(R'' - R\phi'^2) + cR' + \Gamma R - \nu R^3 = 0 ,$$
$$D(R\phi'' + 2R'\phi') + cR\phi' - \omega_2\beta R = 0 , \tag{3.12}$$

where

$$D = \frac{1}{2}(D_1 + D_2), \Gamma = \frac{1}{2}(\alpha_2 + \gamma_2), \quad \nu = \frac{3+\mu}{8}\lambda_0 . \tag{3.13}$$

Note that it is the combinations D and Γ which are the important parameters and not $D_1, D_2, \alpha_2, \gamma_2$ individually. Also, only ϕ' occurs and therefore ϕ is determined to within a constant which is consistent with its interpretation as a phase shift.

We now give qualitative analyses of the special cases with $c \neq 0$ and $c = 0$.

(i) $c \neq 0$: For this case, one can show that bounded solutions can be obtained only if $\phi' = 0$. We then end up with the nonlinear equation for R

$$DR'' + cR' + \Gamma R - \nu R^3 = 0 . \tag{3.14}$$

The qualitative behavior of the solutions can be studied in the (R,R')-plane by setting $T = R'$ and rewriting (3.14) as the first-order system

$$\begin{aligned} R' &= T , \\ T' &= -\frac{c}{D} T - \frac{\Gamma}{D} R + \frac{\nu}{D} R^3 . \end{aligned} \tag{3.15}$$

These equations have three critical points at $(R,T) = (0,0)$ and $(R,T) = (\pm A,0)$ where $A = \sqrt{\Gamma/\nu}$. A local linearized analysis about each critical point shows that the origin is a spiral point if the velocity c lies in the range $0 < |c| < \sqrt{4D\Gamma}$ and an improper node if $\sqrt{4D\Gamma} \leq |c|$; and $(\pm A,0)$ are saddle points.

To illustrate the qualitative features of the solutions, the phase portraits and solutions are sketched in Figures 3 and 4 for $0 < c < \sqrt{4D\Gamma}$ and $\sqrt{4D\Gamma} < c$, respectively. Since we are looking for bounded solutions for all real η the solution trajectories are the separatrices connecting $(\pm A,0)$

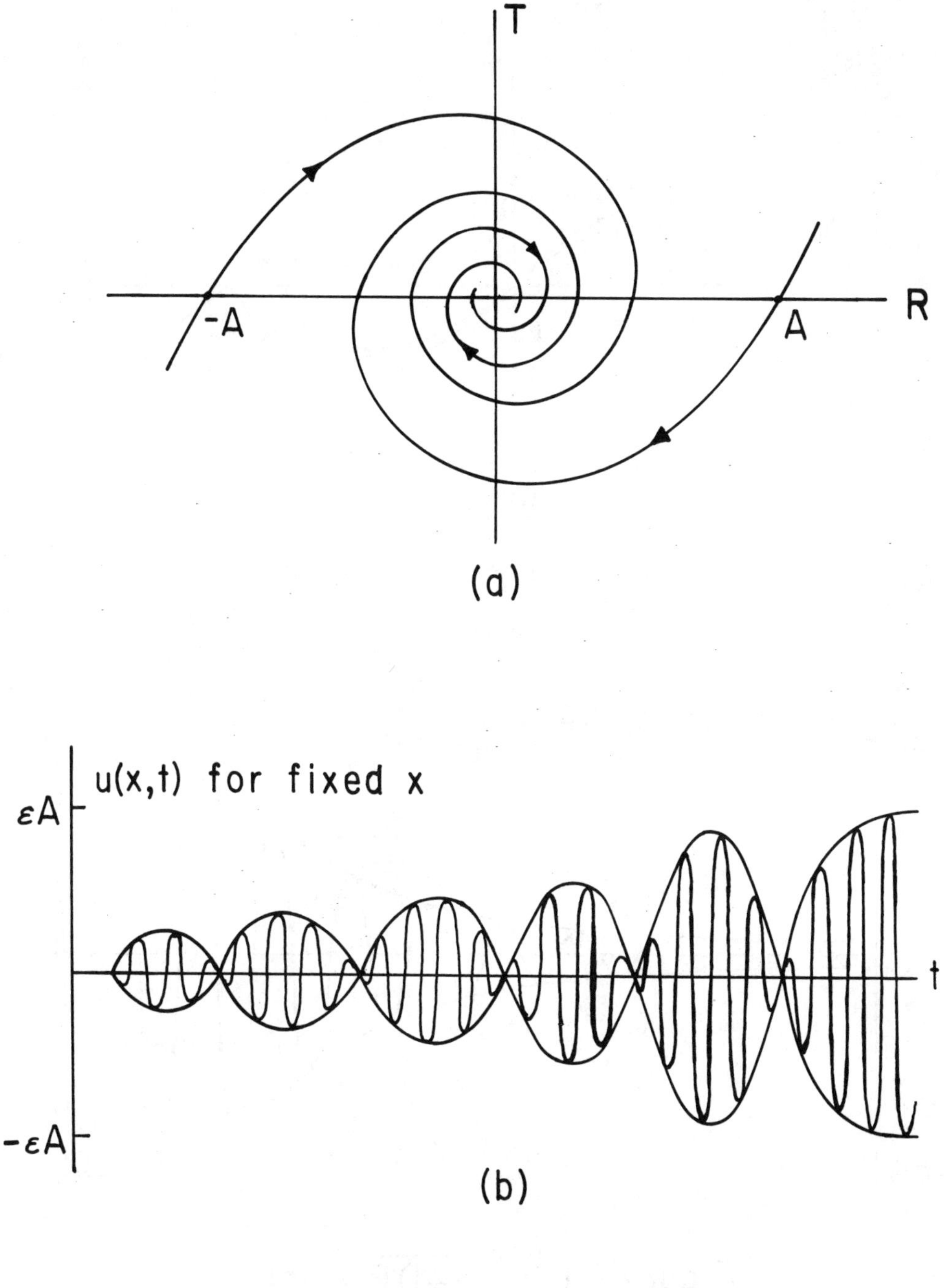

FIGURE 3 – $0 < |c| < \sqrt{4D\Gamma}$

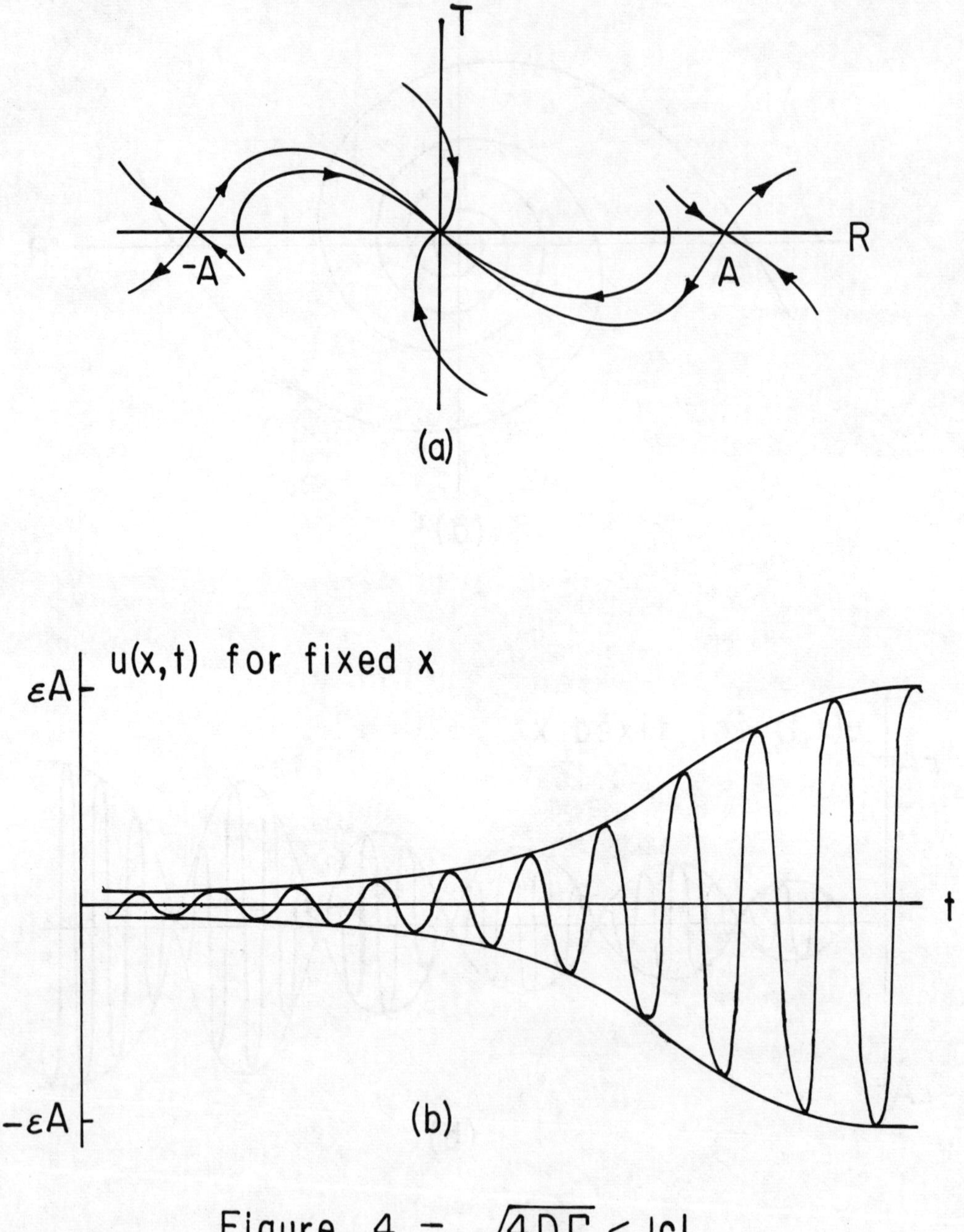

Figure 4 — $\sqrt{4D\Gamma} < |c|$

and the origin. The arrows on the trajectories correspond to decreasing t. It is easy to show that for the system as written, the wave speed c must be nonnegative. However, this is not a real restriction since appropriate changes of variables recover negative values of c.

For $0 < c < \sqrt{4D\Gamma}$, if we stand at a fixed position in x, we see periodically oscillating waves passing by but with oscillating amplitudes (the modulating amplitude). The amplitude oscillation is not periodic. For large negative t this amplitude is small but grows to asymptotically reach εA as $t \to \infty$. For $\sqrt{4D\Gamma} < c$, the amplitude grows monotonically to the asymptotic value εA as $t \to \infty$.

(ii) $c = 0$: In this case, the only interesting nonconstant solutions correspond to the periodic and separatrix trajectories with the origin being a center. Now the solutions oscillate synchronously in time but remain fixed in position. These cases are illustrated in Figure 5 where we note that the abscissa in the solution graphs is x (not t as in Figures 3 and 4). The vertical lines indicate the temporal paths of the solutions.

4. SOME OPEN PROBLEMS

The formal perturbation methods outlined here give oscillatory solutions which are slowly modulated in space and time for partial differential equations of the nonlinear dispersion type or the nonlinear reaction-diffusion type. Two important problems which remain unsolved are to show the solutions are asymptotic and to show how these solutions fit into a more general initial-value problem.

To the order of the approximation carried out, the speed c of the modulating steady progressing wave on the slow space and time scales is not determined. Presumably the wave speed will be determined by the initial data

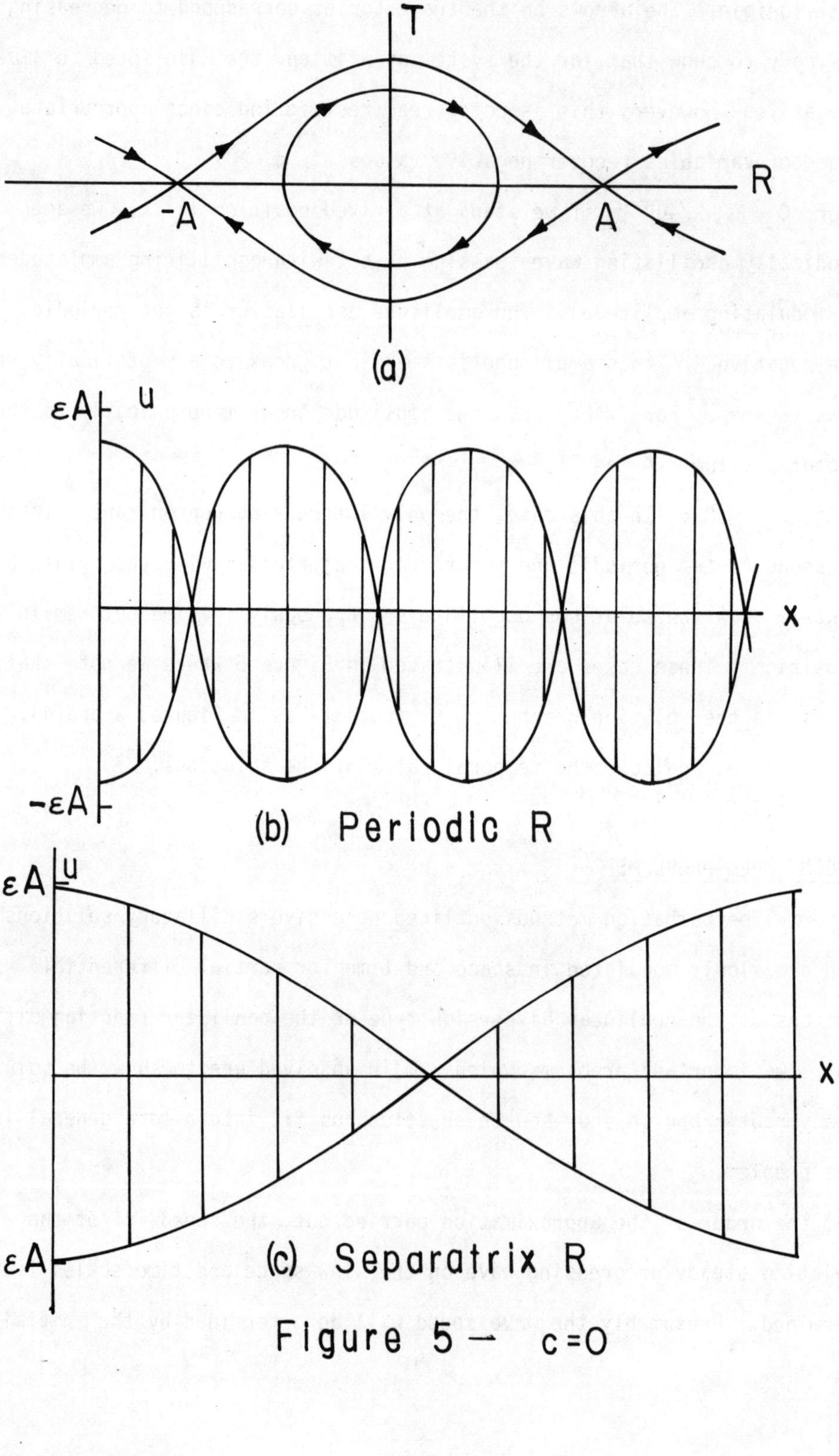

Figure 5 — c=0

although this has not been confirmed for this case. Some determination of the possible range of wave speeds could be determined from a linear stability analysis which remains to be done. It is expected that the stable time dependent solutions would result as asymptotic solutions in t.

In the case when the basic oscillation is not sinusoidal, the above method as given fails since it relies on the use of sinusoidal functions. However, the nonlinear WKB method is suited to this case where the underlying oscillatory behavior would be periodic but nonsinusoidal.

REFERENCES

1. L. N. Howard and N. Kopell, Wave Trains, Shock Fronts, and Transition Layers in Reaction-Diffusion Equations, SIAM-AMS Proc., Vol. 8, 1974, 1-12.
2. L. N. Howard, Bifurcation in Reaction-Diffusion Problems, Advances in Mathematics, 16 (1975), 246-258.
3. L. N. Howard and N. Kopell, Slowly Varying Waves and Shock Structures in Reaction-Diffusion Equations, to be published.
4. M. D. Kruskal and N. J. Zabusky, Progress on the Fermi-Pasta-Ulam Nonlinear String Problem, Princeton Plasma Physics Laboratory Annual Rept. MATT-Q-21 (1963), Princeton, N. J., 301-308.
5. M. D. Kruskal, Asymptotology in Numerical Computation: Progress and Plans on the Fermi-Pasta-Ulam Problem, Proc. IBM Scientific Computing Symposium on Large-Scale Problems in Physics, IBM Data Processing Division, White Plains, N. Y., 1965, 43-62.
6. N. J. Zabusky, A Synergetic Approach to Problems of Nonlinear Dispersive Wave Propagation and Interaction, Proc. Symposium on Nonlinear Partial Differential Equations, W. F. Ames, Ed., Academic Press, New York, 1967, 223-258.

7. G. B. Whitham, Non-linear Dispersive Waves, Proc. Roy. Soc. Ser. A, 283 (1965), 238-261.

8. R. M. Miura and M. D. Kruskal, Application of a Nonlinear WKB Method to the Korteweg-deVries Equation, SIAM J. Appl. Math., 26 (1974), 376-395.

9. N. Kopell, Waves and shocks in an oscillating chemical reaction, this volume.

10. G. Sandri, A New Method of Expansion in Mathematical Physics -- I, Nuovo Cimento Ser. X, 36 (1965), 67-93.

11. D. S. Cohen, F. C. Hoppensteadt and R. M. Miura, Slowly-Modulated Oscillations in Nonlinear Diffusion Processes, SIAM J. Appl. Math., to appear.

Robert M. Miura
Department of Mathematics
University of British Columbia
Vancouver, B. C., Canada V6T 1W5

P NELSON *

Subscriticality for submultiplying steady-state neutron diffusion

ABSTRACT

It is physically reasonable that a system in which each neutron-nucleus collision results in an expected net loss of neutrons must be subcritical (i.e. cannot support a self-sustaining chain reaction). Mathematical proofs of this result, based on the neutron transport equation, have been given; however, corresponding results for the diffusion approximation to the transport equation do not seem to have been studied. In this article we establish a rather weak form of such a result. Possible extensions, both linear and nonlinear, and application to testing of a numerical method are discussed briefly.

1. INTRODUCTION

The distinguishing physical characteristic of neutron transport processes is embodied in the phenomenon of nuclear fission. A fission reaction results, on the average, in a net increase in total neutron population. Such fission processes, with the attendant "chain reactions," are responsible for both the self-sustaining steady flow of energy attained in nuclear reactors and the explosive release of energy from nuclear devices.

A typical mathematical formulation for a time-independent problem, as described by either the transport equation or the diffusion approximation, has the form

$$\phi = K\phi + S . \tag{1.1}$$

*This research was supported in part by NSF Grant No. ENG75-08407.

Here K is a bounded linear operator which is positive in the sense of mapping into itself some cone of nonnegative functions, the "source" term S is nonnegative, and the solution ϕ must be nonnegative in order to be physically meaningful. Let us denote the spectral radius of K by $r(K)$. Relative to the mathematical description (1.1), any particular system then falls into exactly one case of the following trichotomy.

(A) $r(K) < 1$: In this case (1.1) has a nonnegative solution for "arbitrary" nonnegative S. This situation is described by the term subcritical.

(B) $r(K) = 1$: In this event the system is termed critical.

(C) $r(K) > 1$: The term supercritical is used to describe such systems.

Various questions relating to this trichotomy have been studied. Most attention has probably been focused on the critical case. Here one would like, for example, to find sufficient conditions that $r(K)$ be in the point spectrum of K, that there be a nonnegative eigenfunction associated with $r(K)$, and some uniqueness statement for this eigenfunction would also be welcome. These questions clearly call for use of appropriate extensions of the classical Perron-Frobenius-Jentsch results, usually in combination with suitable compactness assertions. Such matters have been studied by Garrett Birkhoff [1-3] for models of the neutron transport equation such that the operator K is "uniformly positive," by Birkhoff and Varga [4] and Habetler and Martino [5] for the finite-difference and continuous-space versions of the multi-group neutron diffusion equation, respectively, and by Victory [6] for the multi-group isotropic-scattering version of the transport equation. Froehlich [7] subsequently studied various extensions of the work of Birkhoff and Varga.

In the supercritical case the usual situation is that there are no nontrivial nonnegative source terms such that a corresponding solution of (1.1) is nonnegative. This reflects the fact that a linear supercritical system cannot support a time-independent neutron population. Consequently it is the time-independent problem which is of dominant interest in this case. Nelson [8] has described the general structure of the nonnegative solutions of (1.1) for S nonnegative and K a compact positive linear integral operator in some L^p-space, $1 \le p < \infty$.

A system is called <u>nonmultiplying</u> (<u>submultiplying</u>) if the expected number of neutrons emerging from a neutron-nucleus collision is not greater than (respectively, less than) unity for any location of the collision or velocity of the colliding neutron. It is physically clear that a submultiplying system cannot support a self-sustaining chain reaction and, with due regard for loss of neutrons by "leakage" from the system, that a nonmultiplying system will ordinarily not support such a reaction. These matters have been studied, for the transport equation and various circumstances, by Olhoeft [9], Case and Zweifel [10], Nelson [11], and by Nelson and Victory [12, 13]. However, no corresponding results for the diffusion approximation seem to have appeared. The primary purpose of this paper is to present an initial attempt to fill this apparent gap. The result obtained is, at best, only partially satisfactory, both in that we are only able to show that submultiplying implies subcritical, and in that the proof of even this weak assertion requires undesirable assumptions of spatial homogeniety. See Section 5 for further discussion of these points.

In Section 2 we introduce notation and describe the precise form of the neutron diffusion equation which is to be considered. Section 3 is devoted to collecting needed properties of the diffusion operator. The proof that submultiplying implies subcritical is given in Section 4. Finally, Section 5 contains a few concluding remarks, including suggestions for possible further study.

2. THE NEUTRON DIFFUSION EQUATION

We write the steady-state neutron diffusion equation in the form

$$-\nabla \; [D(x,E) \; \nabla\phi(x,E)] + \Sigma(E)\phi(x,E)$$
$$= \int_0^\infty k(E,E')\Sigma(E')\phi(x,E')dE' + S(x,E) \qquad (2.1)$$

The independent variables x and E (or E') denote a point in space and neutron energy, respectively. We assume $E \in [0,\infty)$, $x \in \Omega$, where $\Omega \subset E^n$ for some $n = 1, 2$ or 3. The symbol "∇" denotes the gradient operator relative to the spatial variable. The diffusion coefficient and spontaneous source, denoted respectively by D and S, are nonnegative measurable functions defined on $(x,E) \in \Omega \times [0,\infty)$, while the total cross section, Σ, is a positive measurable function defined on $E \in [0,\infty)$. The scattering kernel k is a given nonnegative measurable function defined on $[0,\infty) \times [0,\infty)$. Further assumptions regarding the problem data will be introduced as needed. The unknown function $\phi(x,E)$ represents the scalar neutron flux.

We term the <u>diffusion operator</u> that operator L such that $L\phi(x,E)$ is the left-hand side of (2.1). The first term in $L\phi$ represents the actual diffusion of neutrons, and is an approximation to the streaming term in the neutron transport equation. See the book by Bell and Glasstone [14], and additional works cited therein, for discussion of this approximation. The second term in L represents loss of neutrons from collisions with nuclei of the material comprising the system.

The integral term on the right-hand side of (2.1) represents additions to the scalar flux from neutrons emanating from a collision. This term will be denoted $J\phi$, and J will be referred to as the collision operator. The quantity

$$k_0(E') = \int_{[0,\infty)} k(E,E')dE \tag{2.2}$$

represents the expected number of neutrons emerging from a neutron of energy E' undergoing a collision. The system is nonmultiplying (submultiplying) if $k_0(E')$ is bounded above by 1 (respectively $1 - \varepsilon$, for $\varepsilon > 0$) for $E' \in [0,\infty)$.

The diffusion approximation is usually regarded as best when the flux changes slowly with location. In particular this means it will tend to be a poor approximation in the vicinity of boundaries or material interfaces. Partially because of this, there are no obvious physically correct boundary conditions to be used in conjunction with the diffusion equation. Here we consider (2.1) subject to the Dirichlet-type boundary condition

$$\phi(x,E) = 0, \quad (x,E) \in \partial\Omega \times [0,\infty). \tag{2.3}$$

Several other types of boundary conditions have been proposed and used for the diffusion equation. We refer to Bell and Glasstone [14] (esp. § 3.1e) for a detailed discussion of boundary conditions for the neutron diffusion equation.

3. PROPERTIES OF THE DIFFUSION OPERATOR

In this section we wish to study the equation

$$L\phi(x,E) = f(x,E) , \tag{3.1}$$

where $L\phi$ is defined by the left-hand side of (2.1) and f is a given function. This is simply a partial differential equation in x, with parameter $E \in [0,\infty)$. The classic treatise of Courant and Hilbert [15] (esp.

sections IV.6 and IV.7) is our basic reference for results from the theory of partial differential equations.

We make the basic assumptions that for arbitrary $E \in [0,\infty)$, $D(x,E)$ as a function of x is bounded away from zero and satisfies a Hölder condition with exponent $\alpha = \alpha(E)$, $0 < \alpha < 1$, and that Ω is a bounded domain in E^n such that there exists a strong barrier function at every $x \in \partial\Omega$. (Note the difference in sign between our operator L and that denoted similarly in Ref. 15). Suppose initially that $f(x,E)$ as a function of x is in the space denoted $\hat{C}_\alpha$ in Ref. 15. It then follows that (3.1) subject to the boundary condition (2.3) has a unique solution $\phi = \phi(x,E)$ belonging to $\hat{C}_{2+\alpha}$ as a function of x, for all $E \in [0,\infty)$. Furthermore, the strong form of the maximum principle implies that ϕ is nonnegative if f is nonnegative. If f is nonnegative then, for arbitrary $E \in [0,\infty)$, ϕ must achieve a maximum at an interior point. Equation (3.1) evaluated at this point yields the inequality

$$\Sigma(E) \max\{\phi(x,E):x \in \Omega\} \leq \max\{f(x,E):x \in \Omega\} .$$

By decomposing arbitrary $f \in \hat{C}_\alpha$ into its positive and negative parts, and proceeding essentially as above with f replaced by each of these parts, it is seen that

$$\Sigma(E) \max\{|\phi(x,E)|:x \in \Omega\} \leq \max\{|f(x,E)|:x \in \Omega\}, \qquad (3.2)$$

where $f \in \hat{C}_\alpha$ and ϕ is the associated solution of (3.1) subject to (2.3).

Let B_0 and B_1 denote respectively the normed linear spaces of measurable functions f on $\Omega \times [0,\infty)$ such that the norms

$$||f||_0 = \int_{[0,\infty)} \text{ess sup}\{|f(x,E)|:x \in \Omega\}dE$$

and

$$||f||_1 = \int_{[0,\infty)} \Sigma(E) \text{ ess sup}\{|f(x,E)|:x \in \Omega\}dE$$

are finite. We assert that B_0 and B_1 are Banach spaces under these norms. All properties except completeness are clear. We outline a proof of completeness for B_0. The proof for B_1 is similar.

Let $\{f_n\}$ be a Cauchy sequence in B_0. We claim $f_n(x,E)$ converges in measure relative to Lebesgue measure on $\Omega \times [0,\infty)$. For contrariwise there exists $\varepsilon > 0$ such that for any $N > 0$ there are $m, n \geq N$ such that $A(m,n) = \{(x,E) \in \Omega \times [0,\infty):|f_m(x,E) - f_n(x,E)| > \varepsilon\}$ has measure greater than ε. Let $A(m,n)_E = \{x \in \Omega:(x,E) \in A(m,n)\}$ be the "E-section" of $A(m,n)$. Let μ denote Lebesgue measure, $B(m,n) = \{E \in [0,\infty):\mu A(m,n)_E > 0\}$. Then $\mu A(m,n) > \varepsilon$ implies $\mu B(m,n) > \varepsilon/\mu\Omega$. Therefore for any $N > 0$ there exists $m,n > N$ such that

$$||f_m - f_n||_0 \geq \int_{B(m,n)} \text{ess sup}\{|f_m(x,E) - f_n(x,E)|:x \in \Omega\}dE \geq \varepsilon^2/\mu\Omega .$$

But this contradicts the assumption that $\{f_n\}$ is Cauchy in B_0, and therefore proves $\{f_n\}$ converges in measure. It follows that there exists a subsequence of $\{f_n\}$ which converges, say to f, almost everywhere in $\Omega \times [0,\infty)$. By Fatou's lemma this subsequence also converges to f in B_0, and the fact that $\{f_n\}$ converges to f in B_0 then follows easily from the original assumption that $\{f_n\}$ is Cauchy in B_0.

The collection of $f \in B_0$ such that for each $E \in [0,\infty)$ there exists $\alpha = \alpha(E)$, $0 < \alpha < 1$, with $f \in \hat{C}_\alpha$ is clearly a dense subspace of B_0. According to the above discussion, for each such f there exists a unique function $\phi \in \hat{C}$ for each E such that (3.1) and (2.3) hold, and furthermore the a priori bound (3.2) implies $\phi \in B_1$. We denote the corresponding restriction of L by $\hat{L}$. Thus $\hat{L}$ is a one-to-one operator with domain in

B_1, range a dense subset of B_0, and the inverse of $\hat{L}$ is bounded.

Definition 1. Let G be the closure of $\hat{L}^{-1}$. Then $\phi = Gf$ is the weak solution of the problem (3.1) subject to the boundary conditions (2.3).

The following theorem is an immediate consequence of the preceding discussion.

Theorem 2. For each f in B_0 there exists a unique weak solution in B_1 of the problem (3.1), (2.3). If G denotes the operator taking the data f into the solution ϕ, then G is a bounded linear operator from B_0 to B_1 with norm not greater than unity.

A weak solution is more commonly defined by extending in a certain manner the direct operator $\hat{L}$, rather than the inverse operator as we have done. The next theorem asserts that our concept of weak solution subsumes any such concept based on extending the direct operator.

Theorem 3. Let $\hat{L}'$ be the conjugate operator of $\hat{L}$, denote by B_0' the conjugate space of B_0, and suppose P is a dense subspace of B_0' lying in the domain of $\hat{L}'$ such that $\hat{L}'P$ is (norm-) dense in B_1'. Define the extension $\overline{L}$ of $\hat{L}$ so that $\overline{L}\psi$, $\psi \in B_1$, is the unique element of B_0 such that $h(\overline{L}\psi) = \hat{L}'h(\psi)$ for all $h \in P$. Then $\overline{L}$ is one-to-one, and $\overline{L}^{-1} = G$, where G is as in Theorem 1.

Proof. It suffices to show that $G\overline{L}\psi = \psi$ for all ψ in the domain of $\overline{L}$, and that $\overline{L}Gg = g$ for all g in B_0. Let $\overline{L}\psi = g$, and suppose $g_n = \hat{L}\psi_n$ converges to g in the B_0-norm. Then $Gg_n = G\hat{L}\psi_n = \psi_n \to Gg$, in the B_1-norm. But for $h \in P$ we have $\hat{L}'h(\psi_n) = h(\hat{L}\psi_n) = h(g_n) \to h(g) = \hat{L}'h(\psi)$. As $\hat{L}'P$ is dense in B_2', this shows ψ_n converges weakly to ψ in B_1.

Therefore $\psi = Gg$, which establishes the first desired equality. For the second equality first note that if $g \in B_0$, then we can find a sequence $g_n \to g$ (in the B_0-norm) such that $\{Gg_n\}$ lies in the domain of $\hat{L}$. If $h \in P$ we then have $\hat{L}'h(Gg) = \lim \hat{L}'h(Gg_n) = \lim h(\hat{L}Gg_n) = \lim h(g_n) = h(g)$, which implies $\overline{L}(Gg) = g$. This completes the proof.

4. SUBMULTIPLYING IMPLIES SUBCRITICAL

The following result is proved via simple estimates.

Lemma 4. Suppose $\overline{k} = \text{ess sup}\{k_0(E):E \in [0,\infty)\} < \infty$, where k_0 is defined by (2.2). Then the collision operator, J, is a bounded linear operator on B_1 into B_0 such that $||J|| \leq \overline{k}$.

Definition 5. For $S \in B_0$, a weak solution of the neutron diffusion equation (2.1) subject to the boundary conditions (2.3) is a function $\phi \in B_1$ such that

$$\phi = GJ\phi + GS , \tag{4.1}$$

where J is the collision operator and G is as defined in Section 3.

As both G and J map nonnegative functions into nonnegative functions, (4.1) has the form of (1.1) with K replaced by GJ and S by GS. Furthermore, in the preceding section we showed $||G|| \leq 1$, and for submultiplying systems the inequality $||J|| \leq 1 - \varepsilon$, some $\varepsilon > 0$, is an easy consequence of Lemma 4. Consequently we have the following result.

Theorem 6. Relative to weak solutions of the problem (2.1) and (2.3), submultiplying implies subcritical (i.e. submultiplying implies $r(GJ) \leq ||GJ|| < 1$).

It is worth remarking that the above results easily can be extended to a general setting which encompasses both the continuous-energy case considered above and the multi-group model. One need only replace the energy interval $[0,\infty)$ equipped with Lebesgue measure by an arbitrary σ-finite measure space. The preceding developments can then be carried through mutatis mutandis. (The σ-finite assumption is needed in order to invoke Fubini's theorem in the proof of Lemma 4.)

5. CLOSING REMARKS

The basic motivation for the current work was the desire to find an appropriate basis from which to study the method of variational synthesis as a technique for obtaining approximate solutions to continuous-energy or multi-group neutron diffusion problems. (See Ref. 16, and additional works cited therein, for a description and discussion of variational synthesis relative to such applications.) Variational synthesis has been shown by experiment frequently to give good computational results for such problems, but the situation is complicated by occasional pathology and an absolute dearth of supporting theory. Our original idea was to make an initial attempt to fill the latter void by studying the question of whether variational synthesis would predict subcriticality for nonmultiplying systems. The current work was initiated when it was subsequently realized that the result "nonmultiplying implies subcritical" had not been established for the diffusion equation per se. Theorem 6 above, although actually quite weak, does provide an adequate basis from which to carry out the original plan of studying variational synthesis, especially in view of the fact that we had determined on other bases to limit the initial such studies to the homogeneous case (D, Σ and k independent of x). However, results such as those above seem likely to have some independent interest, and consequently there is reason to consider extensions and related developments. We would like to close with a few comments relative to possible such extension and developments.

Perhaps the most intriguing question is whether results similar to those above can be obtained without requiring Σ and k to be independent of x. The underlying physics gives no clear suggestion as to whether this should be possible. The diffusion approximation is generally regarded as most satisfactory when scattering of neutrons dominates significantly over absorption processes (e.g. capture or fission) and the angular neutron flux is nearly isotropic. The latter condition clearly fails near discontinuities in the data, and therefore it is to be expected that some limitation must be imposed on the spatial variation of the data in order to obtain a satisfactory theory. However, it would be disappointing and somewhat surprising if the restriction required was as stringent as constancy. In order to remove this restriction it does appear that the analysis must be carried out in spaces such that the "x-part" of the norm is defined by something other than the sup-norm. In exploring such possibilities it might be useful to replace the "extrapolated end-point" boundary condition (2.3) (§2.5d of Ref. 14) by the boundary condition of the third kind

$$\phi + 2D\nabla\phi \ n = 0 , \tag{5.1}$$

where n is the outer unit normal to the surface of Ω. Physically (5.1) represents the requirement of no incoming neutrons at the surface of the underlying region.

It would also be of considerable interest to extend the results of Habetler and Martino [5] to the continuous-energy case by showing that $r(GJ)$ is an eigenvalue of GJ with associated nonnegative eigenfunction. The most straightforward approach would be to find conditions under which J is a compact operator from B_1 to B_0, and hence GJ is a compact operator on B_1. This should be fairly easy, although it will require the development of compactness criteria in the space B_0. As GJ is known to be positive, in combination with known results concerning positive operators (e.g. [17]) this would give the desired result. We have not pursued this point here because it is somewhat oblique to our real aim, as described above.

We note that it would be interesting to strengthen Theorem 6 to the assertion "nonmultiplying implies subcritical." A standard sequence of results which would culminate in this conclusion is as follows. First note that Lemma 4 shows that nonmultiplying implies $r(GJ) \le 1$. If one could show that GJ is compact, and then that nonmultiplying implies there cannot be a positive eigenfunction of GJ associated with an eigenvalue of unity, the desired result would follow. It might be easier to establish the latter result under the boundary conditions (5.1) rather than (2.3).

Finally, it seems worth mentioning that a more realistic model of neutron diffusion would incorporate nonlinearities via the dependence of data (cross section, scattering kernel) upon temperature, with the underlying heat source determined by the neutron flux. Stability properties and other questions associated with such nonlinearities have been pursued extensively for the so-called "point kinetics" model, in which spatial dependence is averaged out and the governing equations are a nonlinear system of ordinary differential equations in time; see Chapter 9 of [14] and numerous works cited therein. However, there are interesting phenomena in which the spatial variation is crucial. The recent articles by Keener and Cohen [18] and by Poore [19] concern mathematically interesting aspects of a point-kinetics type reactor model for which the spatial averaging is done separately over the fuel and over moderator and coolant. Initial efforts to study problems with space retained as a continuous variable have been reported by Bronikowski [20] and by Levin and Nohel [21-23], but much remains to be done. It seems likely that many of the techniques described elsewhere in those proceedings could be applied fruitfully to this problem area.

We hope some readers will be stimulated to pursue some of the above suggestions.

REFERENCES

1. G. Birkhoff, Reactor Criticality in Transport Theory, Proc. Nat. Acad. Sci., 45 (1959), pp. 567-569.
2. G. Birkhoff, Positivity and Criticality, Proceedings of Symposia in Applied Mathematica, Vol. 11, Nuclear Reactor Theory, American Mathematical Society, Providence, 1961.
3. G. Birkhoff, Lattices in Applied Mathematics, Proceedings of Symposia in Pure Mathematics, Vol. 2, Lattice Theory, American Mathematical Society, Providence, 1961.
4. G. Birkhoff and R. S. Varga, Reactor Criticality and Non-negative Matrices, SIAM J. App. Math., 6 (1958), pp. 354-377.
5. G. J. Habetler and M. A. Martino, Existence Theorems and Spectral Theory for the Multigroup Diffusion Model, Proceedings of Symposia in Applied Mathematics, American Mathematical Society, Providence, 1961.
6. H. D. Victory, Jr., On the Spectral Radius of Integral Operators Defined by a Class of Difference Kernels, submitted to SIAM J. Math. Anal.
7. R. Froehlich, Positivity Theorems for the Discrete Form of the Multigroup Diffusion Equations, Nucl. Sci. Eng. 34 (1968), pp. 57-66.

8. P. Nelson, The Structure of a Positive Linear Integral Operator, J. London Math. Soc. (2), 8 (1974), pp. 711-718.

9. J. E. Olhoeft, The Doppler Effect for a Non-uniform Temperature Distribution in Reactor Fuel Elements, Report WCAP-2048, University of Michigan, 1962.

10. K. M. Case and P. F. Zweifel, Existence and Uniqueness Theorems for the Neutron Transport Equation,

11. P. Nelson, Subcriticality for Transport of Multiplying Particles in a Slab, J. Math. Anal. Appl., 35 (1971), pp. 90-104.

12. P. Nelson and H. D. Victory, Jr., The Slab Transport Equation in a Space of Measures, Proc. 4th Blacksburg Conference on Transport Theory, to be published by ERDA.

13. P. Nelson and H. D. Victory, Jr., Measures on Phase Space as Solutions of the One-Dimensional Neutron Transport Equation. to appear in Ann. Mat. Pura Appl.

14. G. I. Bell and S. Glasstone, Nuclear Reactor Theory, Van Nostrand Reinhold, New York, 1970.

15. R. Courant and D. Hilbert, Methods of Mathematical Physics, Vol. II, Interscience, New York, 1962

16. P. Nelson, Variational Functionals Which Admit Discontinuous Trial Functions, Nucl. Sci. Eng. 56 (1975), pp. 340-352.

17. S. Karlin, Positive Operators, J. Math. Mech., 8 (1959), pp. 907-937.

18. J. P. Keener and D. S. Cohen, Nonlinear Oscillations in a Reactor with Two Temperature Coefficients, Nuclear Science and Engineering, 56 (1975), pp. 354-359.

19. A. B. Poore, On the Dynamical Behavior of the Two-Temperature Feedback Nuclear Reactor Model, SIAM J. Appl. Math. 30 (1976), pp. 675-686.

20. T. A. Bronikowski, An Integrodifferential System Which Occurs in Reactor Dynamics, Arch. Rational Mech. Anal., 37 (1970), pp. 363-380.

21. J. J. Levin and J. A. Nohel, The Integrodifferential Equations of a Class of Nuclear Reactors with Delayed Neutrons, Arch. Rational Mech. Anal., 31 (1968), pp. 151-172.

22. J. J. Levin and J. A. Nohel, A Nonlinear System of Integrodifferential Equations, in Mathematical Theory of Control, Academic Press, New York (1967), pp. 398-405.

23. J. J. Levin and J. A. Nohel, On a System of Integrodifferential Equations Occurring in Reactor Dynamics, J. Math. and Mech., 9 (1960), pp. 347-368.

Paul Nelson
Department of Mathematics
Texas Tech University
Lubbock, Texas 79409

J RINZEL

Repetitive nerve impulse propagation: numerical results and methods

ABSTRACT

As typical of a variety of nonlinear diffusion phenomena, nerve conduction offers a rich mathematical structure which can include time-independent steady states, traveling waves in the form of fronts, pulses, and periodic trains, along with threshold phenomena. Such structure is frequently difficult to expose analytically yet can be studied computationally using reasonably straightforward numerical methods. This exposition emphasizes the usefulness of the computer as an additional mathematical tool, describes some simple numerical methods, and evidences these objectives with computational results for the FitzHugh-Nagumo nerve conduction equation. First we describe the stimulus-response properties of a uniform nerve model with a steadily maintained stimulus at a fixed location. For an appropriate range of the stimulus intensity, the numerical solution of the partial differential equation corresponds to repetitive firing and propagation of impulses. For large x and t, it has the form of a periodic traveling wave where the frequency depends on the stimulus intensity. We also describe calculations for a non-uniform fiber which illustrate that frequency demultiplication and conduction block can occur when a repetitive train encounters a sudden change in diameter. The periodic traveling waves, observed asymptotically in such signaling problems, satisfy ordinary differential equations in which the propagation speed enters as a parameter. We have also solved these equations numerically and thereby have determined the dispersion relation for the family of periodic wave train solutions. For this we use a direct and simultaneous difference method rather than a shooting scheme. For the numerical solution of the appropriate initial-boundary-value problems which determine the stimulus-response properties we find that straightforward schemes, such as the explicit method or the Crank-Nicolson method, with uniform mesh spacing are frequently adequate. We briefly outline such techniques and discuss some practical pros and cons for them.

1. INTRODUCTION

Theoretical models for nerve conduction exhibit a variety of complex dynamical behaviors. These include traveling waves, time-independent steady states, and time-periodic solutions. Typically such particular solutions are achieved asymptotically for large time and/or distance. Moreover, threshold characteristics with respect to initial and boundary conditions and parameter values are usually associated with the appearance of such solutions. Similar properties are, of course, found for a variety of nonlinear diffusion processes. An introduction to signaling problems for nerve conduction is available in any of a number of review articles (e.g. [7] or [20] and its references). A film prepared by FitzHugh [8] is particularly instructive.

While there are some analytic results for the particular solutions listed above, a good deal of our insight rests on numerical simulations. In this paper we will describe some recent numerical results for the FitzHugh-Nagumo nerve conduction equation. For a uniform axon model we consider the response to a constant stimulating current I applied at a single location. This stimulus is formulated as an inhomogeneous Neumann boundary condition at $x = 0$. For an appropriate range of I values we find repetitive firing of impulses and for large x the numerical solutions appear as periodic wave trains. Associated with this signaling problem are Hopf bifurcations to periodic behavior for critical values of I. We will also describe results in the case where the axon's diameter is different for $x > \overline{x}$ than for $x < \overline{x}$. For the signaling problem, one again finds repetitive generation of impulses just away from $x = 0$ but, if the jump in diameter is too large, these may not be successfully propagated beyond $\overline{x}$. We give an example of frequency demultiplication in which every second impulse fails to propagate beyond $\overline{x}$.

The numerical methods we have employed for these simulations are rather classical. They are fairly straightforward and of course are suitable for rather general diffusion problems. Many readers are no doubt familiar with such techniques and perhaps more efficient ones. Our brief summary of some difference methods for nonlinear diffusion equations will contain little that is new to these readers with, perhaps, the exception of methods for traveling wave solutions. We have included a selection of references and summary on numerics primarily for the benefit of those investigators who might be enticed to adopt the computer as an additional mathematical tool. To implement such methods as we describe, one is not required to be a specialist in numerical analysis nor must one become a programming slave. Of course a user-oriented computer facility which provides readable documentation can be a great asset. Also, an easy-to-use and versatile computer graphics package can be quite helpful to facilitate interpretation and communication of results. On-line capabilities can even add a certain measure of experimental excitement and flavor to theoretical modeling.

This paper is based upon an oral presentation. The intended spirit of that talk was encouragement and motivation for the computational method through illustration of interesting examples from nerve conduction and description of straightforward numerical methods. We admit that the transcription to text of such a pep talk is likely to be less effective and hope that it has not suffered too greatly.

2. REPETITIVE SIGNALING FOR THE FHN EQUATION

Uniform axon.

The FitzHugh-Nagumo (FHN) nerve conduction equation [7, 15] is

$$\frac{\partial v}{\partial t} = \frac{\partial^2 v}{\partial x^2} - f(v) - w$$
$$\frac{\partial w}{\partial t} = \varepsilon(v - \gamma w) \qquad (1)$$

where ε and γ are constants and $f(v)$ is a cubic-shaped function with $f(v) = 0$ for $v = 0, a, 1$ and $f'(0) > 0$. Unless otherwise stated we will take

$$f(v) = v(v - a)(v - 1) \quad . \qquad (2)$$

The model is a conceptual one which exhibits many qualitative features of more quantitatively accountable models such as the Hodgkin-Huxley equation [11]. Here we identify v with potential across the nerve membrane; this is the primary observable in electrophysiology. The auxiliary variable w is usually referred to as the recovery variable. Distance along the nerve is x. The dynamics in (1) can be formally motivated from the type of data which Hodgkin and Huxley analyzed (see [20]). Typical parameter ranges for (1) are $0 < a < \frac{1}{2}$, $0 < \varepsilon << 1$ and $\gamma \geq 0$. The smallness of ε corresponds to the slow rate of recovery processes in nerve membrane. From biophysical considerations, it is reasonable to restrict γ so that $\gamma < 3(1 - a + a^2)^{-1}$. Among other things, this insures that $v = 0, w = 0$ is the unique, uniform, t-independent solution to (1). From our other parameter assumptions, we conclude that this "rest state" is stable.

To briefly motivate the qualitative dynamics for (1) we consider the spatially homogeneous case, $\frac{\partial}{\partial x} = 0$; biophysicists call this the space-clamped preparation. The phase plane profiles for small ε are formally as follows. For an initial voltage displacement $v(0) > a$ $(w(0) = 0)$ the trajectory rapidly carries v,w to the right branch of $w = -f(v)$, the v-nullcline. Then v,w moves slowly up along this branch until it reaches the top. Then it travels rapidly back toward the left branch of $w = -f(v)$ and proceeds slowly down it to make the return to rest. If $v(0) < a$ $(w(0) = 0)$, then v,w returns directly to the left branch and rest without first proceeding along the right branch. Because v,w make a large excursion before their return to rest <u>if</u> $v(0)$ is large enough, the system is said to be excitable and $v = a$ is called the space-clamped voltage threshold. Note, if $\varepsilon = 0$ and $v(0) > a$ then we would have $v \to 1$ as $t \to \infty$. The presence of $w(\varepsilon > 0)$ guarantees the eventual recovery of the rest state; hence w is called a recovery variable.

For the partial differential equation, such a dynamical excursion can be successively translated from one position to the next along the line through the mechanism of diffusion. If a localized region of "nerve" becomes excited then, by passive spread or diffusion, neighboring regions are subsequently recruited. At successive locations, v and w execute an excursion similar to that described above for the space-clamped case. This activity moves as a wave at constant speed along the line and corresponds to the nerve impulse. Experimentally one finds that for a single, steadily propagating, impulse the observed shape and speed are unique to a given nerve. These do not vary with the strength of a stimulus which initiates an impulse.

One way an experimenter might stimulate a nerve is to penetrate it with an electrode and apply a current $I(t)$. If $I(t)$ is too brief or too small in amplitude, there will be no impulse generated; the stimulus must be of adequate strength. If brief, but adequate, stimuli are applied periodically in time, the nerve may "fire" and propagate impulses in a periodic fashion. Similar repetitive firing may occur if I is maintained constant for $t > 0$ and it is large enough; the frequency of impulse firing will vary with the value of I. Neural information is widely thought to be transmitted as some functional of firing frequency.

Analogous to these propagation phenomena, equation (1) has traveling wave solutions. There is a one parameter family of such waves and they are periodic. For such a solution, v and w are periodic functions of $z = kx - \omega t$. For our discussion the period in z is set to one so that k is the reciprocal of pulse spacing or wavelength and ω is the impulse frequency in pulses per unit time. The propogation speed equals ω/k. This one parameter family of periodic wave trains may be described in terms of either k or ω. Such solutions however do not exist for independently specified values of k and ω. The appropriate values satisfy a dispersion relation. Figure 1 illustrates a particular example for the cubic FHN equation. The model parameter values $a = .139$, $\varepsilon = .008$, $\gamma = 2.54$ are only slightly different from those used by FitzHugh [8]. For each ω less than ω_{max} there are two different wave train solutions. The left branch of the curve corresponds to the faster of the two waves. The slow waves (right branch) are thought to be unstable solutions to (1). This has been demonstrated for a particular piecewise linear version of (1) [17, 18]. See also [19] for a general, but formal, argument which implicates ω_{max} as the transition from stability to instability. As $k, \omega \to 0$, we obtain the solitary pulse traveling waves. The fast solitary pulse corresponds to the nerve impulse and the slow one is thought to be unstable (see [6] for more

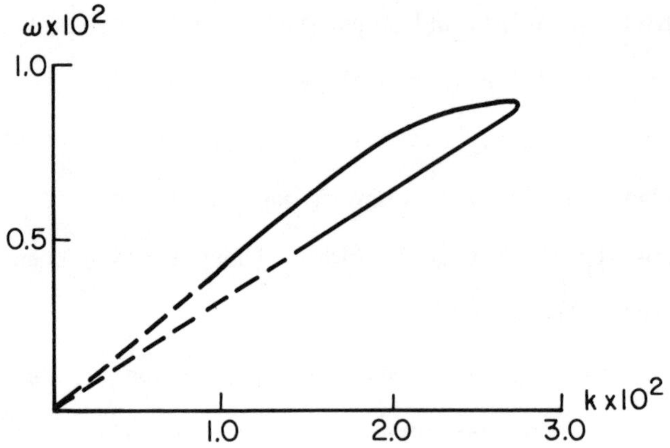

Figure 1
Numerically determined dispersion curve, frequency versus wave number, for periodic wave train solutions to (1). Parameter values in text.

on this). The region of the dispersion curve for small k,ω is shown dashed in Figure 1 because our numerical procedure is not well-suited there and these portions were completed as straight lines through the origin. The slopes of these lines (.4, .32) correspond to the speeds of the fast and slow solitary pulse. For Figure 1, we have $\omega_{max} \doteq .009$ and the speed of this maximum frequency wave is approximately .33, about 15% less than that of the stable solitary pulse. For rigorous proofs on existence of traveling wave solutions to (1) see, e.g. [2], [10]. A singular perturbation construction for $\varepsilon << 1$ of solutions with small k is given in [3].

The family of stable periodic waves corresponds to steady repetitive firing in nerve at different frequencies. Recall that such a response may arise for constant current stimulation. This suggests the following theoretical problem. Consider (1) for $x > 0$, $t > 0$ and with initial conditions

$$v(x,0) = 0, \quad w(x,0) = 0$$

i.e., with the "nerve" initially at rest. The presence of a constant current at $x = 0$ is modeled by the Neumann boundary condition

$$\frac{\partial v(0,t)}{\partial x} = -I/2 \quad , \quad t > 0 \tag{3}$$

where I is constant. Motivated by experimental results we expect to find, for "adequate" values of I, that the solution approaches a periodic traveling wave for large x and t -- repetitive firing. When I is too small experiments indicate that just a few pulses (if any) are generated and then v approaches a time independent steady state. Furthermore, some experiments have shown that if I is too large certain nerves may cease firing repetitively. For intermediate I values the impulse frequency should vary with the value of I.

The initial-boundary-value problem which we have posed has been solved numerically for some values of I for the Hodgkin-Huxley equation [4]. Here we present our numerical results for the FHN equation with the same parameter values used for Figure 1. In our simulations we terminated the axon at $x = L$, $L = 200$, with the following boundary condition

$$\bar{c} \frac{\partial v(L,t)}{\partial x} + \frac{\partial v(L,t)}{\partial t} = 0 \tag{4}$$

where $\bar{c}$ is a positive constant. While there is no particular physiological significance for (4), a rightward traveling wave with speed $\bar{c}$ will satisfy it exactly. Therefore one might argue that this is a justifiable way to treat a semi-infinite axon if traveling waves are anticipated. We set $\bar{c} = 4$, the speed of the stable solitary pulse. Our solutions were computed using a finite difference method, described in the following section, with $\Delta x = .5$, $\Delta t = .25$. Solutions were computed on an IBM 370/168. Results were transferred through a direct linkage to a PDP-10 computer and then displayed at a remote terminal using a three-dimensional graphics package.

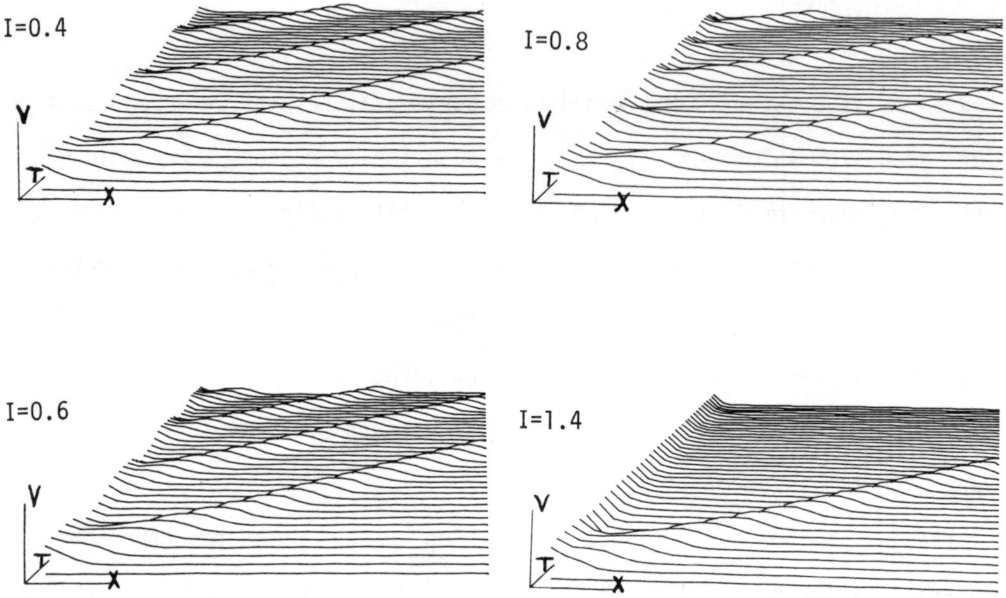

Figure 2

Response, starting from rest, to constant current I applied at $x = 0$. Profiles $v(x,t)$ versus x for discrete values of t from numerical solution of (1), (2), (3) for four values of I.

For four different values of I, Figure 2 illustrates profiles, $v(x,t)$ versus x, for discrete values of t: $t_p = 2.5p$, $0 \leq p \leq 200$. One sees that for $I = .4$, $.6$, $.8$ the solution approaches a periodic wave train for large x. There is an initial transient of a few pulses before the time periodic steady state is reached. The steady or adapted firing frequencies differ for these three cases. The frequency increases from $I = .4$ to $.6$ but then drops sharply between $I = .6$ and $I = .8$. In Figure 3 we have

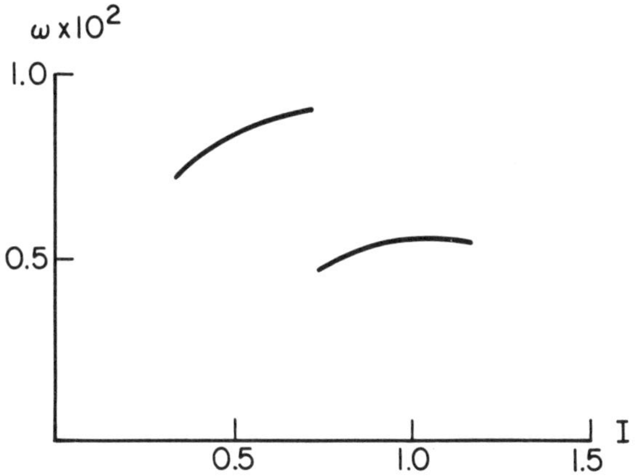

Figure 3
Adapted impulse frequency ω (for large x) as function of current strength I for repetitive firing; see also Figure 2.

plotted firing frequency ω (i.e., reciprocal of the temporal period for a solution which appears to be a traveling wave for large x) versus I for a range of I values. The minimum value of I necessary to illicit repetitive firing here is between .3 and .35. For smaller I the solution tends to a t-independent steady state (not shown in Figure 2). From $I = .35$ to $I = .75$, ω steadily increases; the frequency for $I = .75$ approximately equals ω_{max} for Figure 1. Also over this range of I values, the propagation speed of the observed wave train decreases from $c \doteq .4$ to $c \doteq .33$. Between $I = .75$ and $I = .8$ the frequency suddenly drops in half. Evidently the periodic solution to (1), (3) has lost its stability through a bifurcation to a solution of double the period. Then for I between the approximate values of .8 and 1.15, ω varies continuously although not strictly monotonically. The case $I = .8$ of Figure 2 illustrates that only alternate oscillations near $x = 0$ develop into traveling pulses. If an experimenter records $v(x,t)$ only at $x = 0$, (which is often the case) it is conceivable that this frequency demultiplication phenomenon could go unnoticed, especially if the data $v(0,t)$ does not reveal a noticeable difference in successive oscillations with suspicious likeness in alternate ones. For a short interval of I values above 1.15 (not indicated in Figure 3) we have seen a steady firing pattern of one pulse for each third oscillation at $x = 0$. In a higher I range, which includes the case $I = 1.4$ of Figure 2, the adapted state is apparently time periodic with no pulses, i.e. a standing periodic solution whose amplitude decreases to zero with increasing x. Moreover, the "size" of the oscillation decreases to zero with I; here for "size" we could use $||\Delta v(x) + \Delta w(x)||_\infty$ where $\Delta v(x) = \limsup_{t\to\infty} v(x,t) - \liminf_{t\to\infty} v(x,t)$ and analogously for $\Delta w(x)$. By extrapolation, there appears to be a critical value of I for which the "zero size" limit is achieved. Beyond this, the steady state appears to be t-independent.

Our numerical results suggest that the signaling problem (1), (3) has a stable, time independent solution for small or large I. Analytic results support this suggestion. In general, one can show that there is a unique, time independent (bounded) solution $v(x,t) = \Phi(x;I)$, $w(x,t) = d^{-1}\Phi$ for each value of I [21]. Moreover, for appropriate model parameters, one can expect that there are two critical I values, with $I_* < I^*$, such that $\Phi(x;I)$ is unstable for $I_* < I < I^*$ and stable otherwise. Hopf bifurcation to periodic behavior characterizes the stability transitions at I_* and I^*. These statements have been verified explicitly and in detail for a piecewise linear version of (1) [21]; the general case is being treated with J. Keener. From the numerical results above we conclude, for these parameter values, that the bifurcation is supercritical (soft) at I^* leading to stable small amplitude periodic solutions; the observed frequency matches the theoretically predicted value. However, the sudden appearance of large amplitude periodic solutions for $I \doteq .35$ suggests a subcritical (hard) bifurcation near this lower threshold value.

Nonuniform axon.

The model we have considered until now has assumed that the "nerve" has uniform properties along its length. This idealizes the real situation in which membrane properties and axon diameter may vary with distance. In addition, axons may give off side branches and typically they have a profuse branching structure near their terminal connections with other nerve or muscle cells. Here we will describe numerical results for propagation along an axon model with a simple geometric nonuniformity. We suppose the diameter d of the axon increases abruptly from $d = 1$ for $x < \bar{x}$ to a value $d = \bar{d} > 1$ for $x > \bar{x}$. In general if $d = d(x)$, the appropriate modification to our model is to replace the first of (1) by

$$\frac{\partial v}{\partial t} = (2d)^{-1} \frac{\partial}{\partial x} (d^2 \frac{\partial v}{\partial x}) - f(v) - w . \tag{5}$$

For the case of a diameter jump, the diffusion term has a piecewise constant coefficient and at $\bar{x}$ we have the matching condition

$$\frac{\partial v(\bar{x}_-,t)}{\partial x} = \bar{d}^2 \frac{\partial v(\bar{x}_+,t)}{\partial x} . \tag{6}$$

This is a statement which asserts that current flowing along the core of the nerve is continuous at $\bar{x}$.

We remark that the case of a piecewise constant diameter change for an infinitely long axon is mathematically equivalent to the situation in which an axon branches at $x = \bar{x}$ into n uniform daughter branches each of semi-infinite length and with identical membrane properties but with possibly different diameters d_i . To see this one can scale distance along the i-th daughter branch by $\sqrt{d_i}$:

$$x' = x \quad \text{for} \quad x < \bar{x}$$

$$= d_i^{-\frac{1}{2}}(x - \bar{x}) + \bar{x} \quad \text{for} \quad x > \bar{x} .$$

Then with respect to scaled distance equation (1) applies (after the primes are dropped) for $-\infty < x < \infty$ and gives an identical solution for each branch. The appropriate matching condition at $\bar{x}$ is (6) with $\bar{d}^2$ replaced by $\sum_{i=1}^{n} d_i^{3/2}$ assuming the parent branch has $d = 1$.

To account for impulse propagation, a formal intuitive statement such as the following is often given. The nerve impulse propagates because the advancing front of the pulse provides "adequate" stimulating current for the membrane ahead of it to raise the potential there above threshold. For the FHN equation with $\varepsilon << 1$ and for a uniform nerve model this might be translated into our hypothesis $0 < a < \frac{1}{2}$. Next we follow the extension of such intuitive arguments to the inhomogeneous case. Suppose a localized length of membrane has a higher threshold than the rest of the nerve. If this threshold is sufficiently high or this depressed region is sufficiently long, one could imagine that the pulse might fail to propagate beyond this region. An analogous phenomenon can occur around a region with a local increase in diameter. Here, the area of membrane per unit length which must be brought near threshold has increased. Intuitively, if this increase is too great, an approaching pulse might not provide enough stimulus. In such cases, biophysicists frequently speak of safety factor for successful propagation. The safety factor is said to be low if propagation can be prevented by relatively minor alterations in membrane properties or geometric structure. If rather substantial changes can be tolerated, the safety factor is said to be high. This notion of safety factor is not precisely defined and is invoked in other contexts as well; only rarely is it quantified.

For the following numerical results, the same model parameters apply as in our previous simulations except that here we take $L = 50$ and $\bar{x} = 25$. The terminal boundary condition we use is $\partial v(L,t)/\partial x = 0$ and again we assume resting conditions at $t = 0$. As before, the stimulus is a constant current applied at $x = 0$, i.e. boundary condition (3). For the first case (Figure 4), this stimulus has only finite duration with $I = .3$ for $0 < t \le 20$ and $I = 0$ for $t > 20$. This gives rise to a single impulse which propagates to the right toward $\bar{x}$. In Figure 4A we see that for a diameter increase by a factor of 1.4 ($\bar{d} = 1.4$) the pulse successfully propagates beyond $\bar{x}$. There

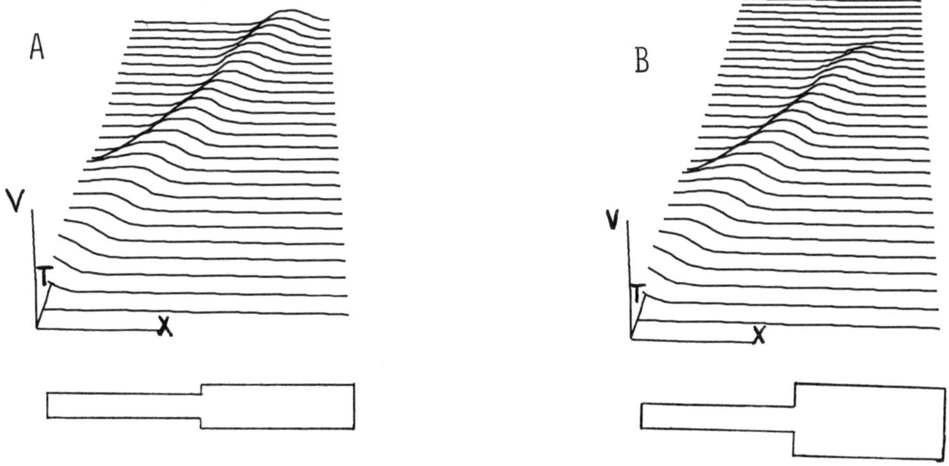

Figure 4

Propagation of single impulse along axon model with jump in diameter by factor $\bar{d}$ at location $\bar{x} = L/2$. Successful propagation beyond $\bar{x}$ for $\bar{d} = 1.4$ (A) and failure for $\bar{d} = 1.45$ (B).

is a transient change in shape as $\bar{x}$ is encountered; the pulse spatial profile for $x >> \bar{x}$ is a stretched version of that for $x << \bar{x}$. Also by a dimensional argument, one concludes that propagation speed for a uniform axon of diameter d is proportional to $\sqrt{d}$. In our case this means the asymptotic propagation speed of a single pulse for $x > \bar{x}$ should be about .47 and numerical results agree with this. The lower plot of Figure 4 illustrates that $\bar{d} = 1.45$ is too great a diameter increase and the pulse does not re-establish itself but dies out.

Next we consider the consequences of a diameter jump for repetitive firing initiated by a maintained current, i.e. constant I for $t > 0$. For Figure 5 A and B, we have $\bar{d} = 1.3$. For this value of $\bar{d}$, a single impulse generated with the nerve initially at rest should successfully propagate through $\bar{x}$. In A we observe successful repetitive firing at frequency $\omega = .0077$ for the case $I = .4$. For a larger stimulus strength, $I = .6$ (as in B), the frequency of pulses generated near $x = 0$ increases to about .0088. Now however we observe that, after an initial transient, only alternate pulses successfully propagate beyond $\bar{x}$ so the frequency of pulses transmitted to large x is .0044. Thus here we see that a single pulse, or successive pulses in a train of moderate frequency, could propagate through $\bar{x}$ but not successive ones in a high frequency train. To understand this we note that each approaching pulse must overcome the hurdle of a change in diameter in addition to the depressing effect of leftover w from the preceding pulse. Hence with a shorter time interval between approaching pulses (compare B to A), not every pulse manages to overcome both of these factors and we have frequency demultiplication of one for two.

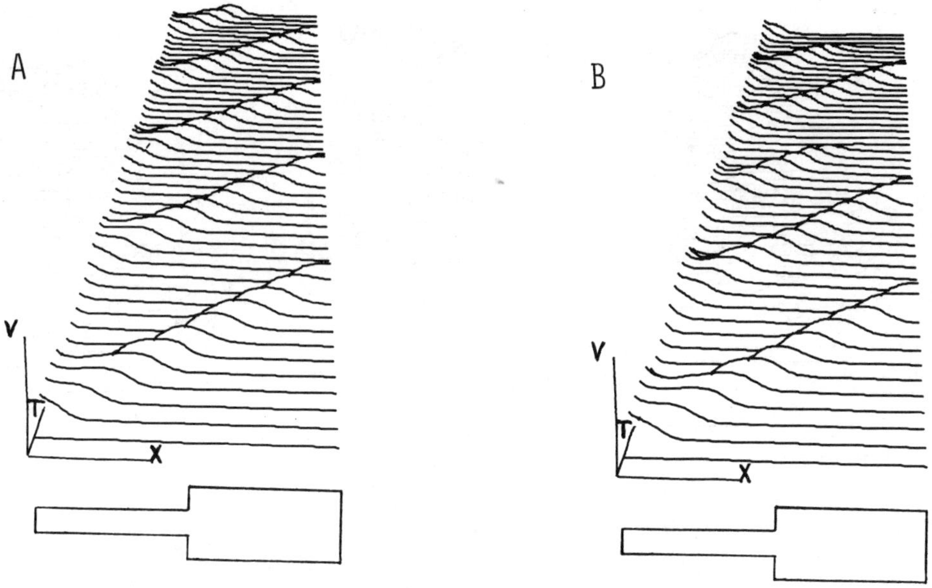

Figure 5

Repetitive firing for axon model with jump in diameter by factor $\overline{d} = 1.3$. Comparison for approaching impulse trains of two different frequencies: $\omega = .0077$ as in A for $I = .4$ and $\omega = .0088$ as in B for $I = .6$. Case B shows frequency demultiplication, only alternate pulses propagate beyond $\overline{x}$.

Finally in Figure 6 we compare three cases of different diameter jumps but each with $I = .6$ so the frequency of approaching pulses is .0088 for each case. In B, $\overline{d} = 1.3$ so that this is the "1 for 2" case of Figure 5B. For a smaller change in diameter, as in A where $\overline{d} = 1.1$, every pulse is successfully transmitted as one might expect. If however $\overline{d} = 1.4$ (case C), we see that only the first pulse succeeds and all the remaining pulses are blocked.

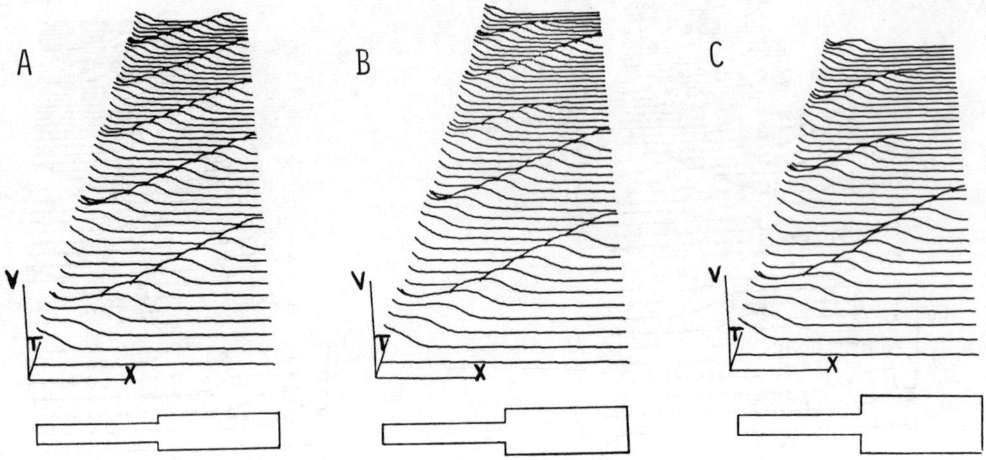

Figure 6

Effect of three different values for $\bar{d}$ on approaching impulse train of given frequency $\omega = .0088$ (I = .6). Successful propagation beyond $\bar{x}$ in A where $\bar{d} = 1.1$. Frequency demultiplication in B for $\bar{d} = 1.3$. One success followed by impulse block for $\bar{d} = 1.4$ in C.

Simulations of nerve conduction in inhomogeneous regions have been carried out by other investigators for different equations; for a review see [13]. In the study of Goldstein and Rall [9] on a three variable model with a step change in diameter, an example illustrates that for a certain range of $\bar{d}$ the approach of a pulse toward $\bar{x}$ is followed by a successfully propagated one beyond $\bar{x}$ and formation of an additional pulse near $\bar{x}$ which propagates back toward $x = 0$. This of course has a different interpretation (see [9]) than the reflected wave one sees from the free end of a vibrating string. In nerve conduction, reflection does not occur at a sealed end where $\partial v/\partial x = 0$; notice the behavior at $x = L$ in Figures 5 and 6. Similarly, by a symmetry argument, impulses which approach each other from opposite directions will annihilate upon collision. The formal argument is that neither pulse can propagate through the opposing pulse's refractory wave; for the FHN equation this corresponds to the w pulse which trails slightly behind the v pulse.

The examples we have discussed are merely illustrative of a variety of possible complex phenomena which may occur in nerve signaling (see [13, 22] for some references to experimental results). For propagation along axons with nonuniform geometry and membrane properties, one should not expect precise maintenance of individual impulse timing during transmission of a pulse train. An additional factor in this regard, even for a uniform fiber, is the dispersive nature of nerve conduction, i.e. the dependence of propagation speed on firing frequency. The quantitative consequences and functional significance of such effects will vary of course, from one situation to the next. Indeed, it is not unreasonable to think that the nervous system may achieve a desired filtering by exploiting such phenomena, e.g. see [22].

The numerical simulations, here and of others, indicate a number of qualitative characteristics shared by a class of nerve conduction models. Such results suggest interesting model problems for which one can hope that additional insight might be obtained analytically.

3. SUMMARY OF NUMERICAL METHODS

Traveling wave solutions.

A traveling wave solution $\phi(kx - \omega t)$, $\psi(kx - \omega t)$ to (1) satisfies

$$\begin{aligned} -\omega\phi' &= k^2\phi'' - f(\phi) - \psi \\ -\omega\psi' &= \varepsilon(\phi - \gamma\psi) . \end{aligned} \qquad (7)$$

It corresponds to a bounded orbit in the phase space for (7). First we will discuss the solitary pulse case. For it, we set $k = 1$ and then ω is identified as the propagation speed c. The problem is to find c such that (7) has a non-constant solution which satisfies $\phi, \psi \to 0$ as $z = x - ct \to \pm\infty$. Typically this problem is approached with a shooting method [7, 11]. By linear stability analysis of $(0,0,0)$ in the (ϕ,ϕ',ψ) phase space, one determines the unique eigenvector direction along which the candidate for a solution approaches rest. A small step in this direction gives a reasonable initial condition from which a straightforward numerical integration (backwards in z) of the equations can proceed. Generally, this candidate is a bounded solution only if the estimate for c is exactly equal to the appropriate value. Hence the calculation is extremely sensitive to this estimate. The solution becomes unbounded in opposite directions depending on whether c is too high or too low. An iterative procedure based upon successive over and under estimates for c frequently leads one to adjustment of c in the last place and even then one seldom gets very far through the pulse.

For a (fast) periodic wave train of large period, the orbit should closely resemble the homoclinic orbit for a solitary pulse. Hence one could expect similar sensitivity characteristics which here include sensitivity with respect to choice of initial conditions. Moreover, since such an orbit passes near the rest point which is hyperbolic, one could expect the orbit to be unstable (as a solution to (7)) in the opposite direction as well. Hence a shooting method in either the forward or reverse direction seems undesirable. Our technique (joint work with R. Miller, details to appear elsewhere) rather treats (7) explicitly as a two point boundary-value-problem in which, for given k, the frequency ω is considered as an eigenvalue parameter. A finite difference approximation is written for each mesh point. Periodic boundary conditions couple the first mesh point to the final one. With appropriate auxiliary conditions the resulting set of algebraic equations are solved iteratively for ω and the periodic solution. There are approximately as many equations as there are mesh points. For another class of reaction-diffusion equations Koppel and Howard [14] have also suggested that such periodic orbits are unstable (in the forward and reverse directions) as solutions to the traveling wave equations and they have developed a different alternative to shooting.

Initial value problem.

By way of introduction and example we consider the linear problem

$$\begin{aligned} &v_t = v_{xx} \quad , \quad 0 < x < 1 \quad , \quad t > 0 \\ &v(0,t) = 0 = v(1,t) \quad , \quad t \geq 0 \\ &v(x,0) = \phi(x) \quad , \quad 0 \leq x \leq 1 \quad . \end{aligned} \tag{8}$$

We will briefly describe simple numerical methods which replace derivatives by finite differences in (8). Let Δx denote the spatial mesh width; for $\Delta x = 1/(J + 1)$ there are J interior mesh points. Let h be the time step. We seek an approximation $u_{j,n}$ to $v(j\Delta x, nh)$.

Perhaps the simplest numerical scheme for (1) is the explicit or forward Cauchy-Euler method given by

$$\begin{aligned} &u_{j,n+1} = u_{j,n} + h\, D^2 u_{j,n} \quad , \quad 1 \leq j \leq J \\ &u_{0,n+1} = 0 = u_{J+1,n+1} \\ &u_{j,0} = \phi(j\Delta x) \quad . \end{aligned} \tag{9}$$

In this formula we have introduced a shorthand notation for the centered second difference operator:

$$D^2 u_{j,n} = (u_{j-1,n} - 2u_{j,n} + u_{j+1,n})/\Delta x^2 .$$

According to this procedure the approximation at the next time level $(n + 1)h$ is given explicitly in terms of that at the current time nh. The method is extremely simple to program for the computer. The primary disadvantage is that a severe restriction on the allowable time step relative to Δx, $h < \Delta x^2/2$, must be satisfied to guarantee stability. If this criterion is met, the error in the solution decreases like h and Δx^2 as the mesh sizes shrink to zero.

For a difference scheme, stability relates to the presence or absence of potential for unbounded growth in solutions to the difference equation. If the difference equation in (9) has a solution which grows with n, one can understand how local errors made in the calculation could be undesirably amplified. Stability then has to do with sensitivity of the difference solution with respect to data which includes round-off error as an inhomogeneous term. For a class of problems which includes (8), stability is checked usually by considering the evolution of Fourier components, i.e. solutions of the form $\lambda_k^n \exp[ik(j\Delta x)]$ where k is an integer and $i = \sqrt{-1}$. Finally we note that stability of the scheme guarantees convergence of the difference solution to the solution of the continuous problem <u>if</u> the difference equation is consistent with the differential equation (i.e., reduces to it as Δx, $h \to 0$). For a thorough discussion of such concepts and the Fourier analysis of stability for a variety of methods for (1), consult [16]; see also [12] as a general reference.

A popular alternative to the explicit method is the Crank-Nicolson scheme. Here the difference expression for v_{xx} is averaged at the two time levels nh, $(n + 1)h$:

$$u_{j,n+1} = u_{j,n} + (h/2)(D^2 u_{j,n} + D^2 u_{j,n+1}) \quad . \tag{10}$$

This is merely an application of the trapezoidal integration rule in the time direction. In contrast to (9) however $u_{j,n+1}$ is no longer explicitly defined. Let us introduce matrix notation with U_{n+1} denoting the vector whose j-th component is $u_{j,n}$. Define the $J \times J$ tridiagonal matrix

$$Q = \begin{pmatrix} -2 & 1 & & & \bigcirc \\ 1 & -2 & 1 & & \\ & \ddots & \ddots & \ddots & \\ & & 1 & -2 & 1 \\ \bigcirc & & & 1 & -2 \end{pmatrix}$$

Now U_{n+1} is implicitly defined as the solution to

$$(I - \kappa Q)U_{n+1} = (I + \kappa Q)U_n \tag{11}$$

where I is the identity matrix and $\kappa = h/(2\Delta x^2)$. While this may at first appear to impose a considerable computational burden, the tridiagonal system (11) can be solved quite efficiently. Explicit solution formulae exist (e.g., see [12] pg. 55) which require on the order of J operations. The main advantage of the Crank-Nicolson method over the explicit method is its stability for all values of h and Δx. With no stability restriction on h one can reduce computational work by taking as large a time step as one's accuracy considerations allow. Because the difference formula is balanced over the two time levels the method is second order accurate in h and Δx. The little additional programming required by the Crank-Nicolson method can be economically worthwhile in the long run. One word of caution may be in order. In the analysis of stability one sees that Fourier components are damped in an oscillatory way and such oscillations can show up in the numerical solution if $h/\Delta x^2$ is too large.

Next let us apply the above methods to a sample nonlinear problem:

$$v_t = v_{xx} + f(v) \tag{12}$$

retaining the same boundary and initial conditions as before. The explicit Cauchy-Euler method extends in a straightforward way to

$$u_{j,n+1} = u_{j,n} + hD^2u_{j,n} + hf(u_{j,n}) \quad , \quad 1 \le j \le J \tag{13}$$

and retains the advantage of programming simplicity. The stability criterion here is the same as that imposed for (9). Moreover, an additional restriction naturally arises here which reflects the dynamics due to $f(v)$. One can intuitively see this by considering $f(v) = -\alpha v$, $\alpha > 0$, for the trivial case of a spatially homogeneous solution to (12). For this, the explicit method requires $2\alpha h < 1$ in order that the difference solution behaves as the continuous solution with $u_n \to 0$ as $n \to \infty$. More generally the explicit method for (12) requires that h be small relative to typical relaxation or growth times $1/|f'(v)|$ imposed by $f(v)$. One, of course, can appreciate that the goal of accurate resolution of transients would itself dictate such a restriction. Numerical solutions for the FitzHugh movie [8] were obtained with the explicit method.

To discuss the Crank-Nicolson method we let $F(U_n)$ denote the vector whose j-th component is $f(u_{j,n})$. Then U_{n+1} is defined implicitly as the solution to the set of nonlinear equations:

$$(I - \kappa Q)U_{n+1} - (h/2)F(U_{n+1}) = (I + \kappa Q)U_n + (h/2)F(U_n). \tag{14}$$

As before, this method has the advantage of unconditional stability. Of course, accuracy considerations as mentioned above may impose a restriction on h. To implement (14) we must give a prescription for solving the non-linear equations. One approach is Newton's method and its advocates claim that one or two iterations is usually adequate. As a first guess U_n is not bad. At each Newton step one solves a linear tridiagonal system with the coefficient matrix $I - \kappa Q - (h/6)\ J_n$ where J_n is a diagonal matrix with elements $f'(u_j^*)$ where u_j^* is the current estimate for $u_{j,n+1}$. The explicit solution formula as previously referenced can be used here. However if J_n has sizable positive elements then h must be restricted to avoid trouble at this stage. As an alternative to Newton's method, successive iteration is sometimes used for (14). For example, at each successive iteration the linear term $(I - \kappa Q)U_{n+1}$ may be treated implicitly to determine the next successive estimate for U_{n+1} while the current estimate replaces the argument of $F(U_{n+1})$. The method is generally easy to implement, but convergence may depend on small enough h. Successive iteration was the choice for the numerical results presented in Section 2; one step of the explicit method provided the first guess for U_{n+1}. A similar method has been used for the Hodgkin-Huxley equations [4].

To avoid solving implicit nonlinear equations, predictor-corrector methods have also been devised for (11) [5]. These methods treat the diffusion term implicitly and do retain stability without an $h/\Delta x^2$ restriction. For each iteration, two tridiagonal linear systems must be solved. Implementation is not particularly difficult. We remark that the successive iteration method outlined above and employed in Section 2 might be viewed as an explicit-predictor/trapezoidal-corrector scheme in which the corrector step may be reiterated a number of times. For yet another class of methods (Galerkin) for solving nonlinear diffusion equations see [1] from this proceedings.

Our discussion has been for a single scalar equation. We hope it is clear how these recipes might be written down for systems of diffusion equations. A problem which frequently arises for multi-component systems is that some components may relax toward equilibrium values much faster than others; these are called stiff systems. Although these components may be changing relatively little after some transient phase their time constants may still be quite short. An explicit method must honor the shortest time constant and thus may impose an inordinately harsh restriction which implicit methods may avoid. On the other hand, such rapid transients may be associated primarily with traveling fronts. If these are present in the spatial domain throughout the integration time, then for accuracy purposes the time step may have to be short anyway. In our model problem (11) or (12), only one type of boundary condition has appeared. A Neumann condition $v_x = 0$ (at $x = 0$ say) can be accurately treated as a symmetry condition. For this, the difference equation is evaluated at $x = 0$ and for $u_{-1,n}$ one uses the value $u_{1,n}$. For the calculations of Section 2, (3) was modeled as an inhomogeneous term $I\,\delta(x)$ in the first of (1) along with even symmetry at $x = 0$.

REFERENCES

1. J. R. Cannon and R. E. Ewing, Galerkin procedures for systems of parabolic partial differential equations related to the transmission of nerve impulses. Proceedings of this conference.

2. G. A. Carpenter, A geometric approach to singular perturbation problems with applications to nerve impulse equations. J. Diff Eqns. (to appear).

3. R. Casten, H. Cohen and P. Lagerstrom, Perturbation analysis of an approximation to Hodgkin-Huxley theory. Quart. Appl. Math. 32, 365-402 (1975).

4. J. W. Cooley and F. A. Dodge, Digital computer solutions for excitation and propagation of the nerve impulse. Biophys. J. 6, 583-599 (1966).

5. J. Douglas and B. F. Jones, On predictor-corrector methods for nonlinear parabolic differential equations. J. SIAM 11, 195-204 (1963).

6. J. W. Evans, Nerve impulse stability. Proceedings of this conference.

7. R. FitzHugh, Mathematical models of excitiation and propagation in nerve, in Biological Engineering (H. P. Schwan, ed.), pp 1-85. McGraw-Hill, Inc., New York, 1969.

8. R. FitzHugh, Impulse propagation in a nerve fiber. National Medical Audio-Visual Center, Atlanta, GA 30333. See also R. FitzHugh, J. Applied Physiol. 25, 628-630 (1968).

9. S. S. Goldstein and W. Rall, Changes of action potential shape and velocity for changing core conductor geometry. Biophys. J. 14, 731-757 (1974).

10. S. P. Hastings, On the existence of homoclinic and periodic orbits for Nagumo's equations. Quart. J. Math., Oxford (to appear).

11. A. L. Hodgkin and A. F. Huxley, A quantitative description of membrane current and its application to conduction and excitation in nerve. J. Physiol. (London) 117, 500-544 (1952).

12. E. Isaacson and H. B. Keller, Analysis of Numerical Methods. Wiley, New York, 1966.

13. B. I. Khodorov and E. N. Timin, Nerve impulse propagation along nonuniform fibres. Prog. Biophys. Molec. Biol. 30, 145-184 (1975).

14. N. Kopell and L. N. Howard, Plane wave solutions to reaction-diffusion equations. Studies in Appl. Math. 52, 291-328 (1973).

15. J. Nagumo, S. Arimoto and S. Yoshizawa, An active pulse transmission line simulating nerve axon. Proc. IRE. 50, 2061-2070 (1962).

16. R. D. Richtmyer and K. W. Morton, Difference Methods for Initial-Value Problems. Interscience (Wiley), New York. 1967 (second edition).

17. J. Rinzel and J. B. Keller, Traveling wave solutions of a nerve conduction equation. Biophys. J. 13, 1313-1337 (1973).

18. J. Rinzel, Spatial stability of traveling wave solutions of a nerve conduction equation. Biophys. J. 15, 975-988 (1975).

19. J. Rinzel, Neutrally stable traveling wave solutions of nerve conduction equations. J. Math. Biol. 2, 205-217 (1975).

20. J. Rinzel, Integration and propagation of neuroelectric signals, in Studies in Mathematical Biology (S. A. Levin, ed.). Math. Assn. of America, Washington, D. C. In press.

21. J. Rinzel, Repetitive activity and Hopf bifurcation under point-stimulation for a simple FitzHugh-Nagumo nerve conduction equation. Preprint.

22. S. G. Waxman, Regional differentiation of the axon: a review with special reference to the concept of the multiplex neuron. Brain Research 47, 269-288 (1972).

John Rinzel
Mathematical Research Branch, NIAMDD
National Institutes of Health
Bethesda, Maryland 20014

M E SCHONBEK *

Some results on the FitzHugh-Nagumo equations

We consider two initial boundary value problems for the system of partial differential equations:

$$\begin{aligned} v_t &= v_{xx} + f(v) - u, \quad x \geq 0, \quad t \geq 0; \\ u_t &= \sigma_v - \gamma u, \qquad\qquad x \geq 0, \quad t \geq 0; \end{aligned} \tag{1.1}$$

where, qualitatively, the graph of f is shown in Figure 1.

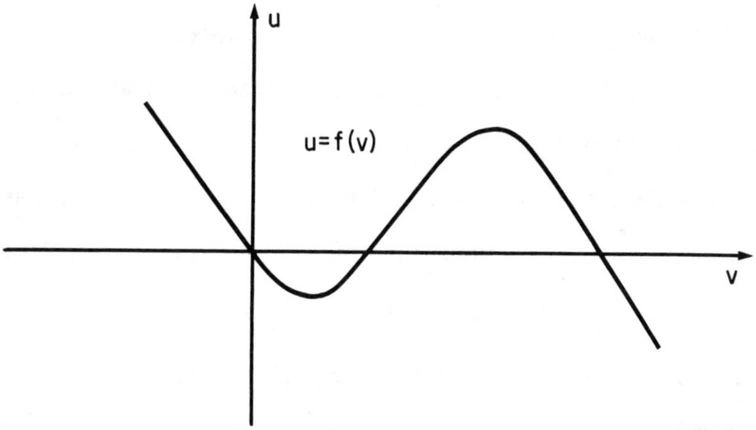

Figure 1

We make the additional assumption that $\sigma/\gamma < -f'(0)$, $\gamma > 0$, $\sigma \geq 0$.

The system (1.1) is an ordinary differential equation coupled to a non-linear diffusion equation. The boundary conditions at $x = 0$ are only given on v. These equations arise as models of the conduction of electrical

*Sponsored by the United States Army under Contract No. DAAG29-75-C-0024 and the National Science Foundation under Grant No. MCS75-17385.

impulses in a nerve axon. The first such model was proposed in 1952 by Hodgkin and Huxley [2]. The form we are using was proposed later by FitzHugh and Nagumo. (See [1] for a discussion of this model.)

We consider several problems. First we treat the case of Dirichlet boundary conditions at $x = 0$. We establish global existence and uniqueness of the solution by the technique of invariant regions of C. Conley and J. Smoller, and the method of contracting rectangles developed by J. Rauch and Smoller in [3].

We then consider the threshold problem for the FitzHugh-Nagumo equations (1.1). Numerical and biological evidence seem to indicate that a strong stimulus of short duration, or a weak stimulus of long duration, is subthreshold. We show that the L_1 norm of the stimulus is a critical parameter. We prove that if the initial data are zero, and the boundary data $v(t,0) = h(t)$ have finite sup norm and vanish outside of some interval $[0,T]$, then our solution is bounded, for all $t \geq 0$, by a constant (which depends on $||h||_\infty$ and T) times the total stimulus, $\int_0^T |h(t)|dt$. Furthermore, we show that if the total stimulus is sufficiently small, the solution has exponential decay. This proves a conjecture of S. P. Hastings [1]. More precisely we prove the following theorems.

<u>Theorem 1</u>. Let $BC(R_+)$ denote the space of bounded continuous functions on R_+, and $C_0(R_+)$ the space of continuous functions on R_+ which tend to zero as x tends to infinity.

Let f be a smooth function which satisfies

1. $f(0) = 0$
2. $-f'(0) > \sigma/\gamma$

3. $\lim\inf|f(v)/v| > \sigma/\gamma$ as $|v| \to \infty$.

Suppose $h \in BC(R_+)$ satisfies

$$h(t) = h(0) = 0 \quad \text{for all} \quad t \geq t_0 \geq 0$$

$$||h||_\infty \leq M \qquad M \text{ positive constant}$$

let $U(t,x) = (v(t,x), u(t,x)) \in C([0,\infty)/C_0(R_+))$ be the unique solution of

$$v_t = v_{xx} + f(v) - u \qquad x \geq 0, \quad t \geq 0$$

$$u_t = \sigma v - \gamma u \qquad x \geq 0, \quad t \geq 0$$

$$v(0,x) = u(0,x) = 0 \qquad x \geq 0$$

$$v(t,0) = h(t) \qquad t \geq 0.$$

Denote by F the vector field $(f(v) - u, \ \sigma v - \gamma u)$. Then for any $T > t_0 \geq 0$ there exists a constant $k = k(T,t_0,M,F)$ such that

$$||U(t,\cdot)||_\infty \leq k||h||_1 \quad \text{for all} \quad t \geq T . \qquad \Box$$

<u>Theorem 2</u>: Suppose conditions 1 to 5 of theorem 1 hold, then there exist constants c, k, and λ, such that if

$$||h||_1 \leq \lambda .$$

Then

$$||U(t,\cdot)||_\infty \leq k \exp(-ct) \qquad t \geq 0$$

where λ and k depend on T, t_0, M and F, and c depends only on F. □

Additional information is obtained by using energy estimates. Suppose that the initial data vanish, the boundary data $h(t)$, are bounded and continuous, vanish outside an interval $[0,T]$ and are sufficiently small. Let α be the first positive zero of $f(v)$. Then by standard multiplier methods we show that if $v(t,x) \leq \alpha - \varepsilon$ for $t \geq T$, $x \geq 0$ and $\varepsilon > 0$, the solution $U = (v,u)$ decays exponentially in $L_2 \cap L_\infty$.

Experiments have not made clear what the correct boundary conditions for (1.1) are. Rauch and Smoller studied the Dirichlet problem in 3 , but no results were known for the Neumann problem. We also prove existence and uniqueness theorems for the FitzHugh-Nagumo equations (1.1) with Neumann boundary conditions at $x = 0$. First we show the existence of a solution for small t. To construct global solutions we obtain an a priori estimate in L_∞ by comparing the solution with the solution $\Phi(t,x) = (\phi(t,x),\alpha(t,x))$ of the "linear FitzHugh-Nagumo equations" $(f(v) = 0)$ with the same initial data and Neumann boundary data. The difference of the two solutions is a function which satisfies equations similar to (1.1), where $f(v)$ is replaced by $g(v) = f(v + \phi)$. Now the initial and boundary conditions are zero. We are then able to apply the technique of contracting rectangles to obtain the desired estimate. More precisely we prove the following theorems:

Theorem 3. Let f be a smooth function which satisfies

1. $f(v) > 0$ if $v < 0$
2. $\lim f(v)/v = -\infty$ as $|v| \to \infty$
3. $\liminf |f(v)/v| > \sigma/\gamma$ as $|v| \to \infty$.

Let $g_1, g_2 \in C_0(R_+)$ and $h \in BC(R_+)$, then there exist a unique solution $U = (v,u) \in C([0,\infty), C_0(R_+))$ to the Neumann problem

$$v_t = v_{xx} + f(v) - u, \qquad x \geq 0, \quad t \geq 0 .$$

$$u_t = \sigma v - \gamma u, \qquad x \geq 0, \quad t \geq 0$$

with initial and boundary data

$$v(0,x) = g_1(x) \qquad x \geq 0$$

$$u(0,x) = g_2(x) \qquad x \geq 0$$

$$v_x(t,0) = h(t) \qquad t \geq 0 .$$

Furthermore for any $T \geq 0$

$$||U(t,\cdot)||_\infty \leq \theta(T, ||h||_\infty, ||g_1||_\infty, ||g_2||_\infty) , \quad 0 \leq t \leq T$$

where θ grows at most exponentially in T. □

We also treat the threshold problem for the FitzHugh-Nagumo equations with Neumann data $v_x(t,0)$. Under the hypotheses analogous to those for the Dirichlet problem (replacing $v(t,0)$ by $v_x(t,0)$) we get, by similar techniques, the same threshold results. In particular, $\int_0^T |v_x(t,0)|dt$ is the critical parameter.

REFERENCES.

1. S. P. Hastings, Some mathematical problems from Neurobiology, American Math. Monthly, 82 (1975), 881-895.
2. A. H. Hodgkin and F. Huxley, A quantitative description of membrane current and its applications to conduction and excitation of nerves, J. Physiol., 117 (1952), 500-544.
3. J. Rauch and J. Smoller, Qualitative theory of the FitzHugh-Nagumo equations, Advances in Mathematics (to appear).

Maria Elena Schonbek
University of Wisconsin, M.R.C.
Madison, Wisconsin 53706

A D SNIDER and D L AKINS

Calculation of transients for some nonlinear diffusion phenomena

In this paper we shall report on some progress made in quantitative analysis of three nonlinear diffusion models. The first two models describe electrochemical processes for which it is desirable to obtain analytical, albeit approximate, expressions for the transient characteristics, so that comparison with experimentally measured curves can lead to estimates for certain chemical rate constants. The third model is concerned with plasma confinement by a toroidal magnetic field; here it will be shown how a modification of the experimental design bypasses, in a sense, the nonlinearity and allows us to predict certain behavior from the linear theory. Actually, the analysis serves to underscore the inadequacy of existing theory in predicting the experimental observations, in this case.

The first electrochemical process we discuss is electrolytically initiated polymerization [1]. Herein a monomer species M is dissolved in a solution into which an electrode is inserted. When the current is turned on, the monomer is polymerized at the electrode

$$M + e \rightarrow P \tag{1}$$

Subsequently the diffusing polymer molecules can either lengthen their chain by attaching another monomer

$$P + M \rightarrow P, \tag{2}$$

or terminate by becoming an inactive species

$$P \rightarrow \text{inactive polymer} \tag{3}$$

We now write the differential equations describing this diffusion-plus-chemical-reaction problem. Denoting the rate constants for (2) and (3) by K_p and K, respectively, and assuming one-dimensional geometry (i.e., a planar electrode), we have the following:

Diffusion and polymerization of M:

$$\frac{\partial M}{\partial t} = \mu \frac{\partial^2 M}{\partial x^2} - K_p MP \tag{4}$$

Diffusion and termination of P:

$$\frac{\partial P}{\partial t} = \rho \frac{\partial^2 P}{\partial x^2} - KP \tag{5}$$

Flux balance at the electrode $(x = 0)$:

$$\mu \frac{\partial M}{\partial x} = -\rho \frac{\partial P}{\partial x} \qquad (x = 0) \tag{6}$$

Nernst's law at the electrode:

$$M(0,t) = \theta P(0,t) \tag{7}$$

(here θ is a constant depending on the electrode potential and certain activity coefficients; see [2]).

Initial values of concentration:

$$M(x,0) = C \tag{8}$$

$$P(x,0) = 0 \tag{9}$$

The boundary conditions at the container wall are discussed in [1], but here we shall just consider the semiinfinite-container case.

Equations (4) - (9) describe a diffusion process with a nonlinear term appearing in (4) only. The solution procedure in [1] is a straightforward expansion of M and P in powers of K_p ;

$$M(x,t) = \sum_{\ell=0}^{\infty} M_\ell(x,t)K_p^\ell \tag{10}$$

$$P(x,t) = \sum_{\ell=0}^{\infty} P_\ell(x,t)K_p^\ell \tag{11}$$

Using a Green's function derived in [1] one obtains recursion relations for the Laplace transforms of M and P ; there the inverse transforms are derived for the first order terms. The physically measureably quantity is the electrode current $i(t)$, which is related to flux by

$$i(t) = -\eta\rho \frac{\partial P}{\partial x}(0,t) \tag{12}$$

where η is another constant [1]. Using the power series for P, one finds

$$i(t) = \frac{\eta\sqrt{\mu}\ C}{(1+\sqrt{\mu/\rho}\ \theta)\sqrt{\pi t}} \{1 + (\frac{K\theta}{\sqrt{\rho/\mu} + \theta} - 2K_p\ C\sqrt{\rho/\mu}\ \beta)t + \ldots\} \tag{13}$$

This is a very useful expression; by plotting $t^{1/2}i(t)$ versus time for two different initial concentrations C and comparing the initial slopes, one can determine the values of the two unknown constants K and K_p . This was the ultimate goal of the analysis.

The second process which we analyze is transient electrochemiluminescence (ecl) [3,4]. The system consists of benzoyl peroxide as the bulk oxident precursor, tetraethylammonium perchlorate as the supporting electrolyte, and benzonitrite as the solvent; for the convenience of the non-specialist, we shall subsequently employ symbols for the various species without naming them explicitly (see [5]). The process evolves as follows:

Initially there are uniform concentrations of species R and O. When the electrode $(x = 0)$ is turned on $(at\ t = 0)$, these species are reduced at the electrode to species R^- and ϕ^- , respectively. The latter species diffuse back into the solution and initiate a sequence of reactions involving by-products ϕ, R_3, and R^+. The species R^+ interacts with R^- to produce light, whose intensity is measured as a function of time. The goal of the analysis is to use this measured curve to predict the values of the various rate constants governing the reactions.

Previous authors [6] have used digital simulation or curve-fitting with exponentials to predict these constants. However, the former technique only leads to compositive terms coupling rate constants with certain efficiencies, while the latter ignores the effects of the diffusion processes. We shall describe an approximate, "semi-analytic" approach to the problem which overcomes both objections.

We begin by listing the reactions governing <u>ecl</u> [4].

At the electrode:

$$R + e^- \rightarrow R^- \qquad (14)$$

$$O + 2e^- \rightarrow 2\phi^- \qquad (15)$$

In the solution:

$$R^- + O \rightarrow R + \phi^- + \phi \qquad (16)$$

$$R^- + \phi \rightarrow R_3 + \phi^- \qquad (17)$$

$$R_3 + O \rightarrow R^+ + \phi^- + \phi \qquad (18)$$

$$R^+ + R^- \rightarrow 2R + \text{radiation} \qquad (19)$$

The approximations which we introduce are motivated by considering the relative importance of these reactions in the chain of steps which produces R^+, the "fuel" for the radiation mechanism (19). This "hierarchy analysis" is derived from the following description.

Initially there is a large background concentration of O and R, and no chemical activity occuring. The electrode is pulsed, producing R^-, which diffuses back into the solution. (Also ϕ^- is produced, but it does not enter into the <u>ecl</u> mechanism.) Nothing happens at a given spatial point until a significant level of R^- builds up; R^- then reacts with the background concentration of 0 to produce three species: ϕ^-, which is unimportant, R, which slightly reinforces the background concentration, and ϕ. After the latter is produced it reacts locally with the diffusing R^- to give R_3, and more ϕ^-. The R_3 which is generated then reacts with the background 0 to give more ϕ (a second-generation reproduction), more ϕ^-, and R^+. Finally, the R^+ annihilates with the diffusing R^- to yield photons.

This picture lends creditability to the following specific assumptions, which we use to analyze the transient behavior:

(i) The (background) concentrations of 0 and R are constant (for $x > 0$) in space and time. (The initial concentrations are so large that one can ignore their depletion by the higher-order reactions and decomposition at the electrode surface.)

(ii) The concentration of R^- is determined by the production at the electrode, the diffusion process, and the loss mechanism of reaction (16). (All other intermediate species occur in much smaller concentrations.)

(iii) The intermediate species ϕ^-, R_3, and R^+ do not diffuse, but react locally. Reaction (18) can be ignored (in comparison to reaction (16)) as a production mechanism for ϕ^-.

As a result, the differential equations governing the various concentrations for this "hierarchy model" are

$$\frac{\partial C_{R^-}}{\partial t} = D \frac{\partial^2 C_{R^-}}{\partial x^2} - k_1 C_0 C_{R^-} \tag{20}$$

$$\frac{\partial C_0}{\partial t} = k_1 C_0 C_{R^-} - k_2 C_{R^-} C_\phi \tag{21}$$

$$\frac{\partial C_{R_3}}{\partial t} = k_2 C_{R^-} C_\phi - k_3 C_0 C_{R_3} \tag{22}$$

$$\frac{\partial C_{R^+}}{\partial t} = k_3 C_0 C_{R_3} - k_4 C_{R^+} C_{R^-} \tag{23}$$

$$C_0 = \text{constant} \tag{24}$$

$$C_R = \text{constant} \tag{25}$$

where k_1, k_2, k_3, and k_4 are rate constants for reactions (16), (17), (18), and (19) respectively. Initially, the concentrations of R^-, ϕ, R_3, and R^+ are everywhere zero except at the origin, where we have

$$C_{R^-}(0,t) = C_R \; ; \tag{26}$$

this is the "diffusion-limiting" form of the flux-balance law and Nerst law.

Now equation (20) is the only partial differential equation, and using the Green's function mentioned earlier we find

$$\begin{aligned} C_{R^-}(x,t) = C_R[1 &- \exp(-k_1 C_0 t)\, \mathrm{erf}(x/2\sqrt{Dt}) \\ &- k_1 C_0 \int_0^t \exp(-k_1 C_0 \tau)\, \mathrm{erf}(x/2\sqrt{D\tau}) d\tau] \end{aligned} \tag{27}$$

("erf" denotes the error function).

Substituting (27) we can successively solve the <u>linear</u>, <u>ordinary</u> differential equations (21), (22), (23) to obtain an integral expression for $C_{R^+}(x,t)$. Unfortunately, the formula is still untractable and we need a simple approximation for $C_{R^-}(x,t)$. Observe that if the product $k_1 C_0 t$ is small, C_{R^-} is nearly proportional to the complementary error function, $\mathrm{erfc}(x/2\sqrt{Dt})$. Furthermore, an examination of the graph of $\mathrm{erfc}(\lambda)$ reveals that

$$\begin{aligned} &\mathrm{erfc}(\lambda) < .25 \quad \text{if} \quad \lambda > .8 \\ &\mathrm{erfc}(\lambda) > .75 \quad \text{if} \quad \lambda < .2 \quad . \end{aligned} \tag{28}$$

This suggests approximating erfc by a Heaviside function with its discontinuity located at $\lambda = .5$. As a pointwise approximation, this substitution is outrageous; but since C_{R^-} only occurs under integral signs in the formulas, it is a mean-square approximation that we need, and this is not a bad L_2 approximation. Thus we propose

$$C_{R^-}(x,t) = \begin{cases} 0 & \text{if} \quad t < x^2/D \\ C_{R^-}(x,\infty) & \text{if} \quad t > x^2/D \, . \end{cases} \tag{29}$$

Physically, an observe at $x > 0$ sees a "wall" of R^- propagating toward him at a speed $\sqrt{D/t}$; when it passes, C_{R^-} immediately rises to its saturation value $C_{R^-}(x,\infty)$, which, by (27), is

$$C_{R^-}(x,\infty) = C_R \exp(-x\sqrt{k_1 \, C_0/D}) \, . \tag{30}$$

With the hierarchy approximation for the chemical reactions and the Heaviside approximation for the diffusion mechanism we are able to obtain a closed-form expression for

$$\int_0^\infty C_{R^+}(x,t) \, C_{R^-}(x,t)dx \; ,$$

which is proportional to the radiation intensity $I(t)$. The formula looks like (see [5] for details)

$$I(t) \propto A_1[1 - \exp(-\sqrt{k_1 C_0 t})] + A_2 \mathcal{D}(\sqrt{k_2 C_R t}) + A_3 \mathcal{D}(\sqrt{k_3 C_0 t}) + A_4 \mathcal{D}(\sqrt{k_4 C_R t}) \quad (31)$$

where $\mathcal{D}(\lambda)$ is Dawson's integral

$$\mathcal{D}(\lambda) = \exp(-\lambda^2) \int_0^\lambda \exp(\rho^2)d\rho \quad . \quad (32)$$

The authors have had good success [5] in fitting experimental data by using a least squares algorithm on (31) to determine the rate constants k_1, k_2, k_3, k_4. Thus it appears that the approximations are justified and can be used in other chemical-diffusion models. It is of interest to note that keeping track of the diffusion mechanism in the mathematics dictates fitting the intensity curve with Dawson integrals, and represents a refinement of Faulkner's suggestion [6] of fitting with exponentials.

The final model which we discuss here is entirely unrelated to the others; it analyzes the molecular diffusion of a magnetically confined plasma. To see the full implication of the experimental results it will be convenient to review the kinetic theory of diffusion.

When a beam of moving particles enters a volume which is populated by obstacles, the fraction of the beam which is scattered out after penetrating a distance dx is given by [7]

$$\frac{\text{(density of scatterers)(volume of target)(cross section of each scatterer)}}{\text{(cross section of target)}}$$

This leads to the formulae

$$\text{mean free path} \equiv r_m = 1/n_\sigma \sigma \quad (33)$$

$$\text{mean time between collisions} \equiv \tau = r_m/v = 1/n_\sigma v\sigma \quad (34)$$

where

n_σ = density of scatterers

σ = cross section of each scatterer

v = speed of beam particle

and certain statistical averages are implied in (33) and (34). Kinetic theory then shows that this process leads to Fick's law (the "diffusion equation") for the density of particles in the beam, n;

$$\frac{\partial n}{\partial t} = \vec{\nabla} \cdot (D\vec{\nabla} n) \tag{35}$$

Here the diffusion coefficient D has the interpretation [7,8]

$$D = \frac{(\text{average distance traveled between collisions})^2}{(\text{average time between collisions})} \tag{36}$$

In an isotropic situation, the average distance travelled between collisions is the mean free path r_m, so that

$$D = \frac{n_\sigma v\sigma}{n_\sigma^2\sigma^2} \propto \frac{1}{n_\sigma} \tag{37}$$

However, for a magnetically confined plasma, the charged particles gyrate around the fields, so on the average the shift in a particle's trajectory would be equal to this radius of gyration, the "Larmor" radius [7], so that (at least for the case of diffusion perpendicular to the field lines)

$$D = \frac{r_L^{\ 2}}{\tau} = \frac{e^2B^2n_\sigma v\sigma}{m^2} \propto n_\sigma \quad . \tag{38}$$

For the modern stabilized toroidal devices ("Tokamaks") a further correction must be considered. Here the lower-energy particles follow closed orbits whose projections onto a cross-section plane, perpendicular to the field, is banana shaped [9]. For such particles the average shift in trajectory due to a collision is the "banana width"

$$D = \frac{r_B{}^2}{\tau} = \frac{n_\sigma v\sigma m V_{toroidal}}{e\ B_{poloidal}} \propto n_\sigma \tag{39}$$

At any rate, (38) and (39) both argue that the diffusion coefficient is proportional to the density of scatterers,

$$D = \alpha n_\sigma \tag{40}$$

It is well known that for linear diffusion problems the "confinement time" , i.e. the time required for the depletion of a significant fraction of the concentration, is inversely proportional to D. More precisely, if we set

$$D(\vec{r}) = D_0 d(\vec{r}) \tag{41}$$

where D_0 is a scale factor and $d(\vec{r})$ is a "profile function" whose maximum value is 1, then one can solve (35) by separation of variables as follows: if

$$n(\vec{r},t) = R(\vec{r})T(t)\ , \tag{42}$$

then

$$\dot{T} = \lambda D_0 T \tag{43}$$

and

$$d(\vec{r})\nabla^2 R = \lambda R \tag{44}$$

Permissible values λ_n of the eigenvalue λ are determined by the geometry for equation (44), and the associated time dependence is

$$T_n(t) = e^{\lambda_n D_0 t} = e^{-t/\tau_n}\ ; \tag{45}$$

each τ_n - in particular, the confinement time for the principal made τ_1 - varies as $1/D_0$.

However, for the plasma containment problem a nonlinearity occurs because the scatterers are the same species as the "scatterees," i.e., the beam scatters itself. Hence $n_\sigma \equiv n$ and (35) becomes, with (40),

$$\frac{\partial n}{\partial t} = \vec{\nabla} \cdot (\alpha n \vec{\nabla} n) \quad . \tag{46}$$

If we try separation of variables here, (42) leads to

$$\dot{T} = \lambda T^2 \tag{47}$$

and

$$\vec{\nabla} \cdot (\alpha R \vec{\nabla} R) = \lambda R \tag{48}$$

The solution of (47) is

$$T(t) = \frac{-1}{\lambda(\tau_c + t)} \tag{49}$$

indicating an inverse power, rather than exponential, law of decay, but since superposition fails it is difficult to see how (49) fits into the scheme of things.

In an attempt to study how the containment time varies with density n, experimenters constructed a design which operates the Tokamak in a steady-state mode by adding a (constant) source. To analyze this, observe that the confinement times for the principal mode in (45) and for the "separated" mode in (49) both obey the relation

$$\tau_c = \frac{\text{initial density}}{\text{initial rate of decay of density}} \tag{50}$$

Hence, if the device is operated in a steady-state mode with a source, the confinement time can be estimated by the "replenishment time"

$$\tau_r = \frac{\text{density (constant)}}{\text{replenishment rate (constant)}} \tag{51}$$

("Proof": if Q represents the source term, then the steady state diffusion law reads

$$\vec{\nabla} \cdot (D\vec{\nabla}n) = Q \; . \tag{52}$$

Now turn off the source: the law becomes

$$\vec{\nabla} \cdot (D\vec{\nabla}n) = \frac{\partial n}{\partial t} \tag{53}$$

Hence if n remains spatially continuous,

$$\frac{\partial n}{\partial t} = Q \tag{54}$$

and the right-hand sides of (50) and (51) are equal. Thus $\tau_c = \tau_r$ is validated, to the extent that the equations (50) and (51) are true.)

Let us now assume that, in the "pumped" plasma, the steady-state density $n(\vec{r})$ is expressed as a product of N_0, the peak density, and a "unit profile function" $\mu(\vec{r})$,

$$n(\vec{r}) = N_0 \, \mu(\vec{r})$$

The replenishment rate (per unit volume) equals the divergence of the flux of n ; hence by Fick's law

$$\text{replenishment rate} = \vec{\nabla} \cdot (D\vec{\nabla}n) = N_0^{\,2}\vec{\nabla} \cdot (\alpha\vec{\nabla}\mu) \; .$$

Consequently the replenishment time varies inversely with density,

$$\tau_r = \frac{N_0\mu(\vec{r})}{N_0^{\,2}\vec{\nabla} \cdot (\alpha\vec{\nabla}\mu)} \propto \frac{1}{N_0} \tag{55}$$

for devices with similar profiles.

This conclusion was checked against experimental data by Dr. R. Parker, Director of the National Magnet Laboratory, M.I.T. He collected data from the various Tokamak projects and plotted the results as in Figure 1 [10].

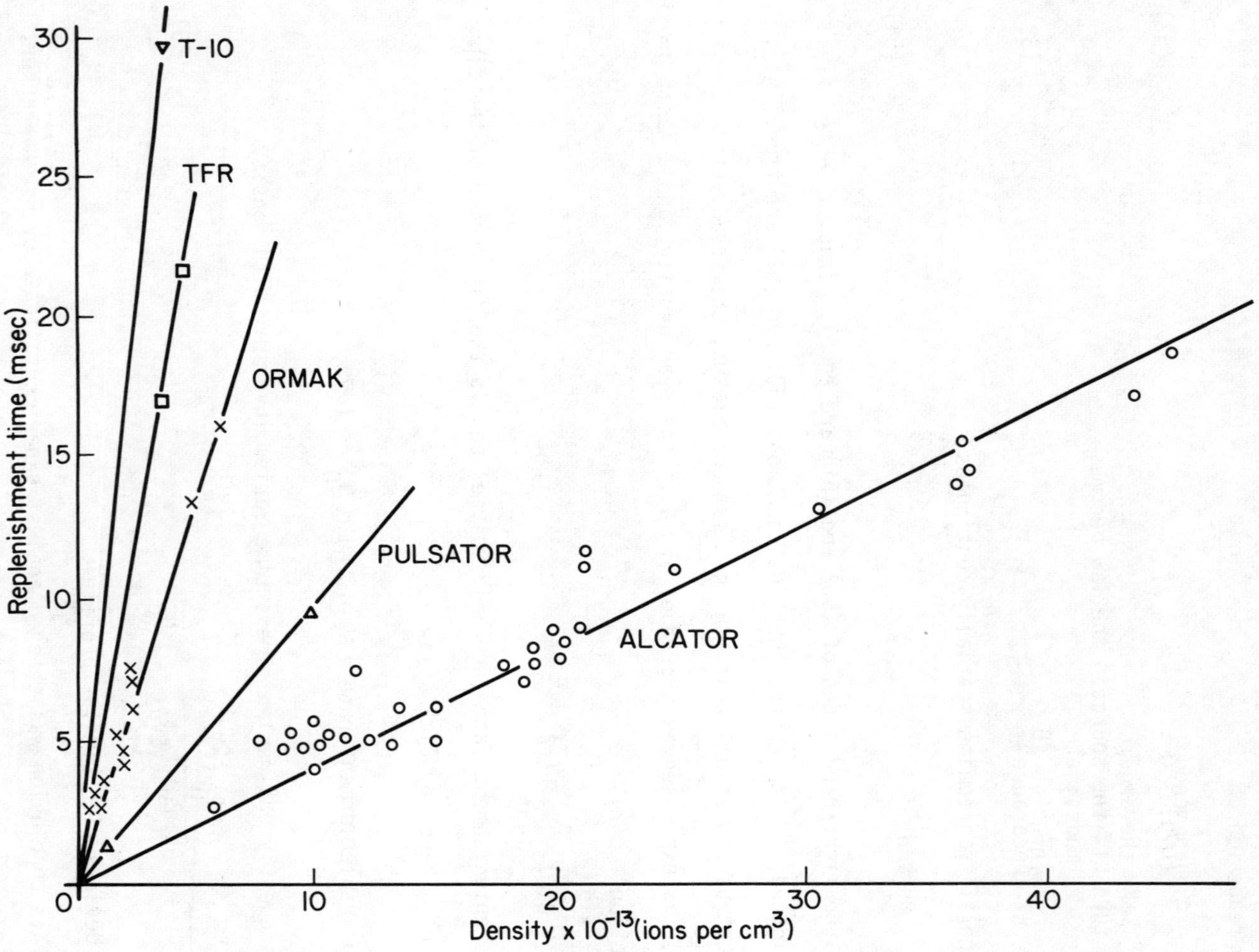

Fig. 1: replenishment time vs. Density for various experiments

The conclusion is unavoidable - τ_r varies directly with density, in contradiction to (55).

This phenomenon is at present an open question in theoretical physics. Certainly the arguments leading to (55) are not rigorous (we have ignored mixing of modes, etc.), and we would expect there to be deviations from (55), but the appearance of a well-documented, simple, experimental law in such drastic disagreement with theory is quite perplexing.

REFERENCES

1. A. D. Snider, D. L. Akins and R. L. Birke, Transient Analysis of Electrolytically Initiated Polymerization, to appear in J. Electroanal. Chem.
2. D. G. Gray and J. A. Harrison, Polymerization of an Electroactive Species at the Rotating Disc Electrode, J. Electroanal. Chem., 24, 187 (1970).
3. D. L. Akins and R. L. Birke, Energy Transfer in Reactions of Electrogenerated Aromatic Anions and Benzoyl Peroxide. Chemiluminescence and its Mechanism, Chem. Phys. Letters, 29, 428 (1974).
4. T. D. Santa Cruz, D. L. Akins and R. L. Birke, Chemiluminescence and Energy Transfer in Systems of Electrogenerated Aromatic Anions and Benzoyl Peroxide, J. Amer. Chem. Soc., 98, 1677 (1976).
5. D. L. Akins and A. D. Snider, Transient Analysis of Electrochemiluminescence, submitted to J. Chem. Phys.
6. L. R. Faulkner, A New Approach to the Analysis of Chemiluminescence Transients from Step Experiments, J. Electrochem. Soc., 122, 1190 (1975).
7. F. F. Chen, Introduction to Plasma Physics, Plenum Press, N. Y., (1974).

8. J. Crank, The Mathematics of Diffusion, 2nd ed., Clarendon Press, Oxford (1975).

9. A. A. Galeev, Neoclassical Theory of Transport Processes, in Advances in Plasma Physics, v. 5, J. Wiley and Sons, N. Y. 1974.

10. C. B. Stevens, Current Status of Controlled Fusion Energy Research, January 1977, Executive Intelligence Review, Campaigner Publications, N. Y. (1977).

Arthur David Snider
Department of Mathematics
University of South Florida
Tampa, Florida 33620

Daniel L. Akins
Department of Chemistry
University of South Florida
Tampa, Florida 33620